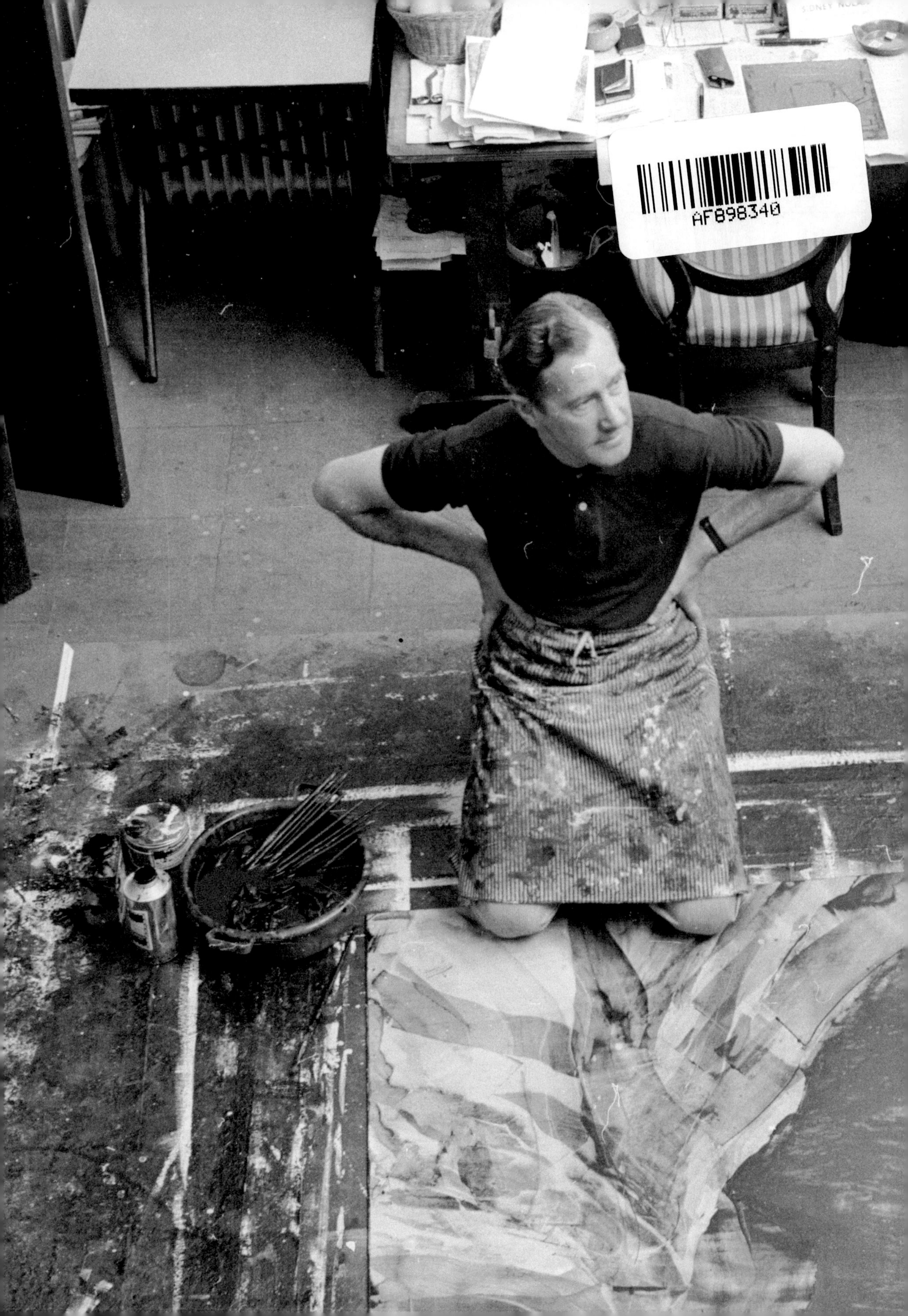

Other titles in the
Artist's Materials series

Willem de Kooning

Lucio Fontana

Sam Francis

Hans Hofmann

Jean Paul Riopelle

Paula Dredge

THE ARTIST'S MATERIALS

Sidney Nolan

with contributions by
Anne Carter and Gillian Osmond

The Getty Conservation Institute
Los Angeles

The Getty Conservation Institute

The Getty Conservation Institute (GCI) works internationally to advance conservation practice in the visual arts—broadly interpreted to include objects, collections, architecture, and sites. The Institute serves the conservation community through scientific research, education and training, field projects, and the dissemination of information. In all its endeavors, the GCI creates and delivers knowledge that contributes to the conservation of the world's cultural heritage.

Published by the Getty Conservation Institute, Los Angeles
Getty Publications
1200 Getty Center Drive, Suite 500
Los Angeles, California 90049-1682
www.getty.edu/publications

Beatrice Hohenegger, *Project Editor*
Anne Canright, *Manuscript Editor*
Jeffrey Cohen, *Designer*
Suzanne Watson, *Production*
Kelly Peyton, *Image and Rights Acquisition*

Distributed in the United States and Canada by the University of Chicago Press
Distributed outside the United States and Canada by Yale University Press, London

Printed in China

Library of Congress Cataloging-in-Publication Data
Names: Dredge, Paula, author. | Carter, Anne (Conservator), contributor. | Osmond, Gillian, contributor. | Getty Conservation Institute, issuing body.
Title: Sidney Nolan : the artist's materials / Paula Dredge ; with a contribution by Anne Carter and Gillian Osmond.
Other titles: Artist's materials.
Description: Los Angeles : The Getty Conservation Institute, [2019] | Series: The artist's materials | Includes bibliographical references and index.
Identifiers: LCCN 2019015939 | ISBN 9781606065945 (pbk.)
Subjects: LCSH: Nolan, Sidney, 1917–1992. | Painting, Modern—20th century—Technique. | Paint materials—Analysis. | Artists' materials—Analysis.
Classification: LCC ND1105.N6 D74 2019 | DDC 759.994—dc23 LC record available at https://lccn.loc.gov/2019015939

FRONT COVER: Sidney Nolan, *Self Portrait*, 1943 (detail, fig. 25)
PAGE i: Gerti Deutsch, Sidney Nolan in his St. John's Wood studio, London, 1960 (detail, fig. 76)
BACK COVER: Albert Tucker, Sidney Nolan in the Loft, Parkville, 1945.

Contents

Foreword

Throughout his career, Australian artist Sidney Nolan (1917–1992) experimented with non-traditional materials and application methods, ranging from used cardboard and Masonite, commercial and household paints, additives and dryers, crayon wax and spray paint. His narrative style of figurative painting introduced historical figures and the Australian landscape to new audiences. His most famous work, the Ned Kelly series, comprises twenty-seven paintings portraying Australia's most illustrious outlaw and folk hero.

We're pleased to share *Sidney Nolan: The Artist's Materials*, the first comprehensive study of Nolan's methods and use of unconventional materials. Apprenticed as a sign painter at the age of thirteen, his early experience with commercial paints made him one of the first artists of his generation to ultimately use only non-artists' paints, later giving him a distinct advantage over other artists who were just beginning to use these new materials in their own works.

This is the sixth book in the Getty Conservation Institute's (GCI) Artist's Materials series, which grew out of the institute's Modern and Contemporary Art Research Initiative. The initiative explores the new and varied materials used by artists of the period and shares this knowledge with conservation professionals who are charged with, and challenged by, caring for these works. Previous books in the series cover Willem de Kooning, Lucio Fontana, Sam Francis, Hans Hofmann, and Jean Paul Riopelle.

Paula Dredge, head of Paintings Conservation at the Art Gallery of New South Wales and the author of the present book, first came to the attention of the GCI via her extensive body of research on the paints, including many pots of Ripolin that were left in Nolan's studio after his death. Ripolin is a brand of house paint that is perhaps much better known in Europe than Australia and that has been used by several highly influential artists, including Pablo Picasso and Francis Picabia. The use of house paints and other industrial coatings by modern and contemporary artists has long been of interest to the GCI as part of its Modern Paints project, which develops conservation strategies for the very different surfaces produced by industrial materials compared to artists' tube colors.

The results of Paula's research, which include a scientific examination of more than thirty of Nolan's paintings and an analytical and historical study of the materials left in the artist's studios in Wales and Sydney, coupled with our own interest in the paint pots, brought us to the realization that Nolan would be an excellent candidate for a book in our Artist's Materials series.

We thank Paula for sharing her exploration of Nolan's career, unique vision, artwork, and experimental working methods. She has been a generous colleague, and we are most grateful for her dedication to this important study and volume.

Our thanks also go to Anne Carter and Gillian Osmond from the Queensland Art Museum, who generously contributed to the chapter on Nolan's travels through Queensland, and to the many individuals and institutions who helped make Paula's study and *Sidney Nolan: The Artist's Materials* possible.

Timothy P. Whalen
John E. and Louise Bryson Director,
Getty Conservation Institute

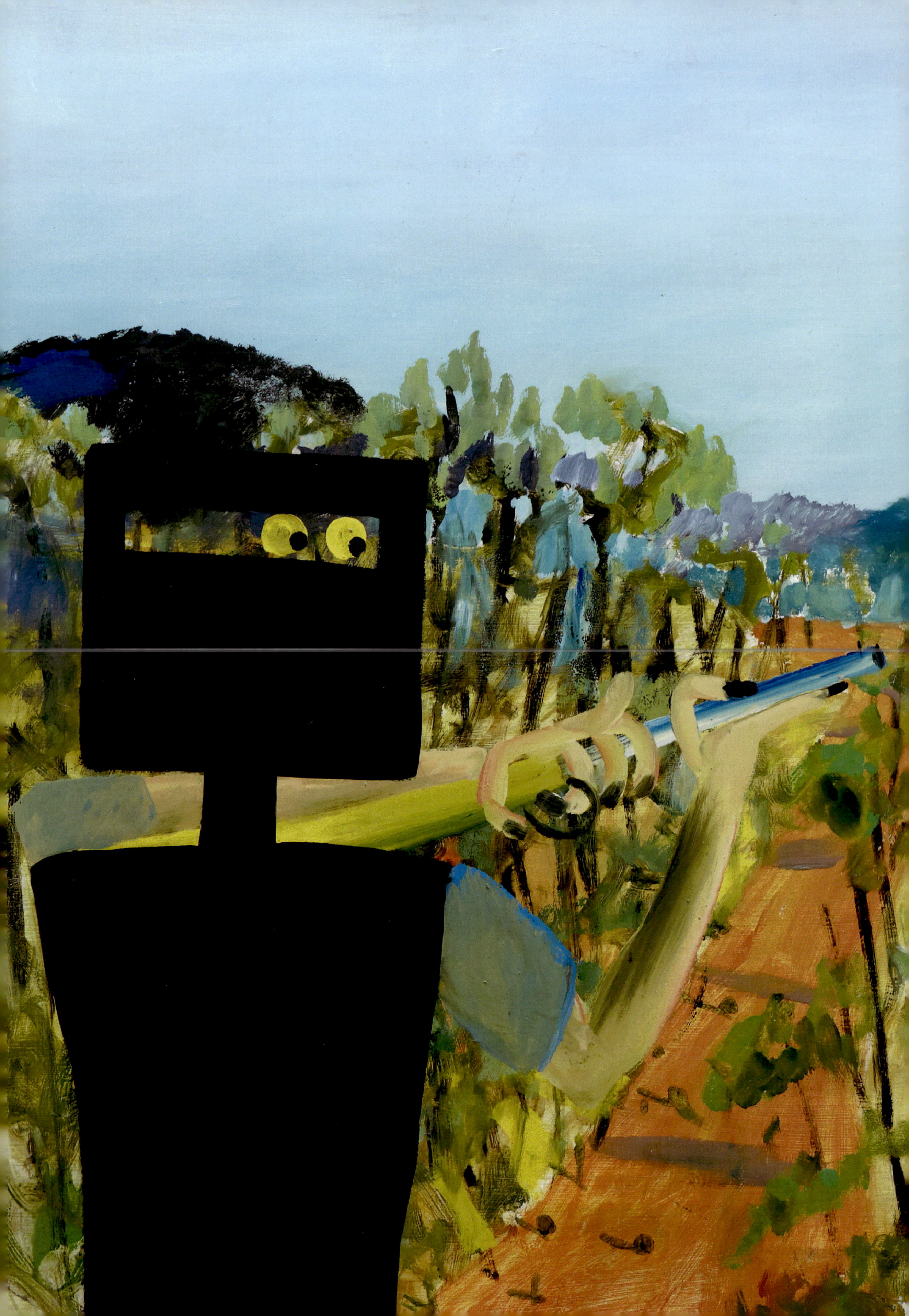

Introduction

Sir Sidney Nolan (1917–1992) is one of the most experimental modern painters to emerge from the austerity of the 1930s and the violence of the Second World War. Painting with glossy and intensely colored commercial paints on scraps of cardboard, blotting paper, and sheets of builder's wallboard called Masonite®,[1] he invented a new Australian identity through evocative landscapes and storytelling. Nolan often painted a troubled European presence, using the subjects of nineteenth-century historical figures—the bushranger Ned Kelly, the shipwreck survivor Eliza Fraser, the central Australian explorers Robert O'Hara Burke and William John Wills—and such contemporary subjects as drought-affected cattle. He continues to be celebrated for his ability to make the people and places he painted extraordinary.

Nolan was born into a working-class family in Melbourne and grew up through the interwar years. He undertook training in the trade of commercial art at Prahran Technical College and started work in a factory at the age of fourteen painting illuminated signs for the roofs of commercial cars. From 1933 to 1937 he worked in the advertising department of Fayrefield Hats, a men's felt hat company, cutting out shapes and painting window and trade displays. Despite having little, if any, training in the fine arts, on the eve of the Second World War Nolan decided to give up paid employment and embark on a career as an artist.[2] The works painted between 1938 and 1947 are some of the most inventive and seminal of his career. Few, however, sold at their first exhibition, and it was not until the late 1940s, when he was settled and exhibiting in Sydney, that he found critical and commercial success.

Financial reward and the end of the Second World War gave Nolan his first opportunity to travel outside Australia. In 1950 he drove through Spain, Portugal, Italy, and France in a Hillman Minx car freighted from Australia. Three years later he moved permanently to England, though he continued his travels, including living for nine months on the Greek island of Hydra. Beginning in 1958 he spent several years in the United States, with extensive travels in Africa in 1962, including Kenya, Tanzania, and Ethiopia, following in the footsteps of his favorite poet, Arthur Rimbaud (1854–1891). He traveled to Antarctica in 1964, and then Mexico, India, China, and Cambodia. Between his travels Nolan lived in London, where he had a successful career through the 1960s and '70s. In the 1980s Nolan moved to the Welsh border country and in 1983 purchased an estate called The Rodd. His last decade at The Rodd was remarkably creative, the results of which are only now becoming known and appreciated.

Nolan fearlessly applied his experience and knowledge of non-artist's paints from commercial industry to his own art in the early years of the Second World War. Although it was not uncommon for artists emerging in the postwar decades to explore materials

Detail of Sidney Nolan, *First-Class Marksman*, 1946. See fig. 30.

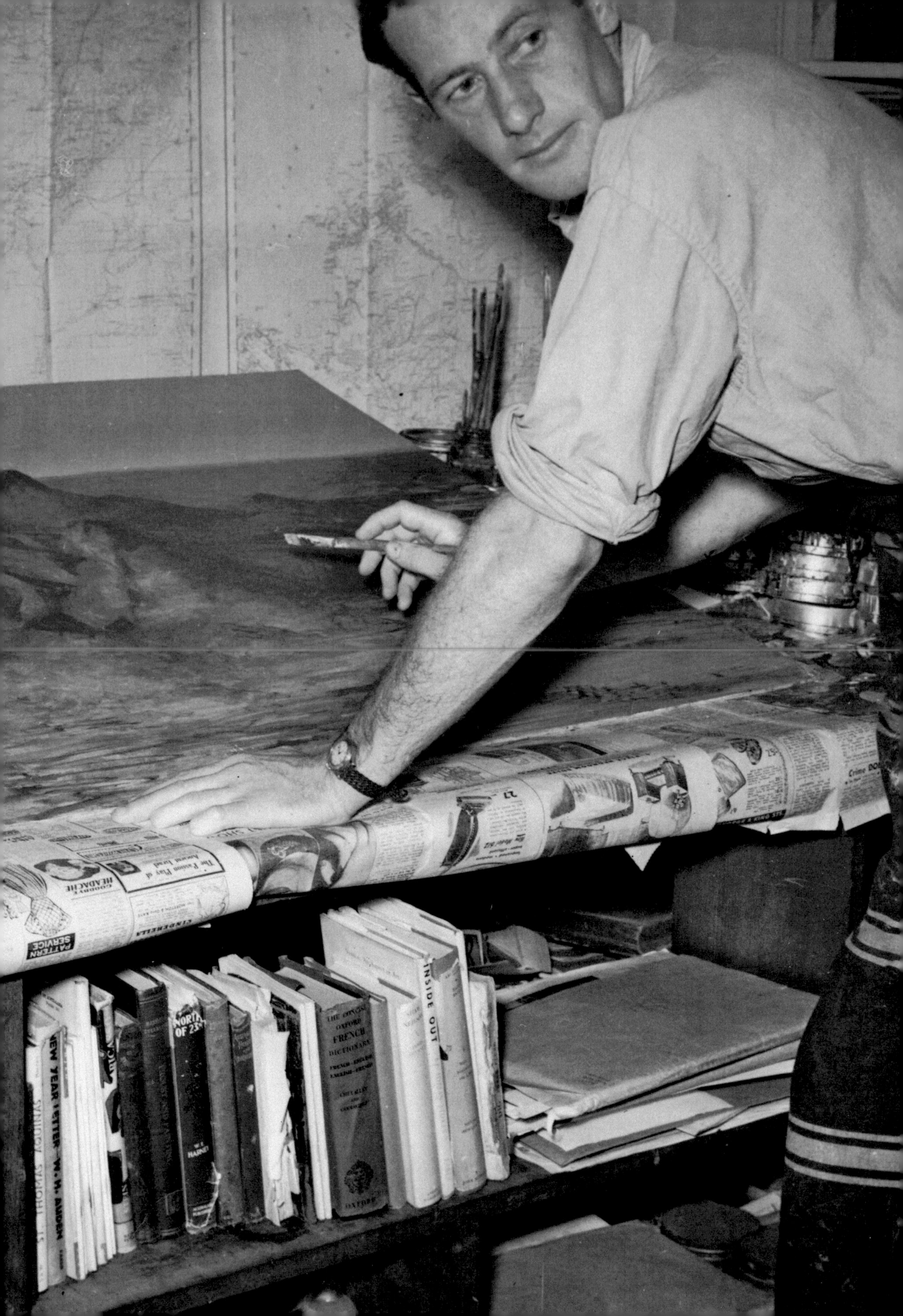
PATTERN SERVICE
NEW YEAR LETTER—W. H. AUDEN
THE CONCISE OXFORD FRENCH DICTIONARY
OXFORD
INSIDE OUT

(right)
Brian Adams, Nolan painting at the Rodd, 1986.

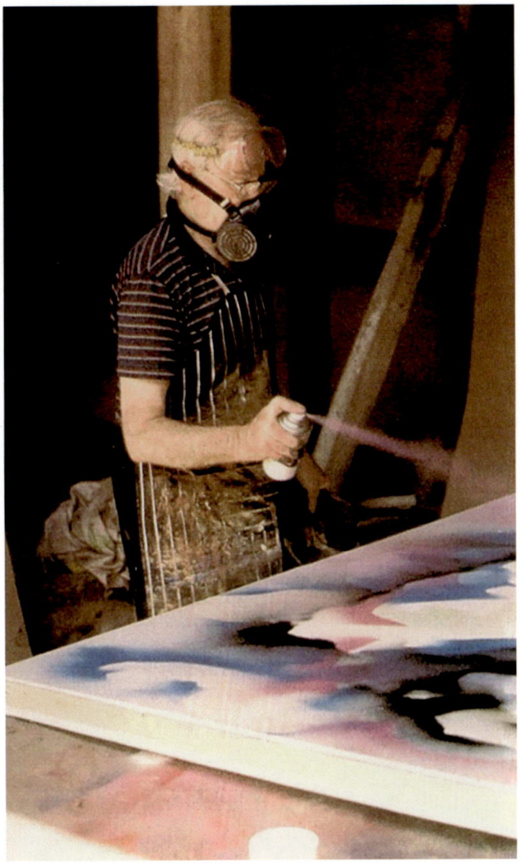

they had come to know while working in paint factories, house painting, sign making, and commercial art, such applications were neither common nor accepted practice for artists in the 1930s and '40s. While many artists may have adopted these more widely available paints as a short-term strategy during the austere war years, Nolan went on to form a lifelong attachment to new and emerging materials. Wartime supply issues, moreover, created an artist who ever after stockpiled his supplies, making Nolan's materials a bountiful subject for study still today. He left behind two studios: one in Sydney, predating his move to England in 1953, and the other at his estate The Rodd, which contains more than a thousand items preserved at his death in 1992.

Travels to new places often occurred alongside changes in Nolan's materials or techniques. He was at the artistic vanguard in his adoption of any new synthetic product he could turn to his purposes for inventive and fast painting. Nolan's technical and material progression thus parallels developments in paint technology. In this regard, a study of his materials can provide a useful guide to similar materials used by other artists. Numerous types of materials were synthesized in the interval between the development of nitrocellulose into a paint in the 1920s and the introduction of acrylic and vinyl emulsion paints in the 1950s. While the paint industry itself struggled to develop consistent terminology for these fast-changing formulations, the end user, whether house painter or artist, would have found it even more difficult to untangle the confusion of brand names, trade terminology, and scientific nomenclature. Nolan, however, was extremely knowledgeable about his materials, seeking out and interrogating paint manufacturers and chemists to obtain not only emerging materials from the production floor but also information about their formulations and properties.

(left)
A. L. Frazer, Nolan painting *Unnamed Ridge*, 1949. See fig. 68.

Many other artists tried to emulate the success they saw emerging from Nolan's cans of house paint, particularly Ripolin®, but they struggled to control the liquid flow and problems of miscibility—problems that Nolan relished and even emphasized. The technical challenge of a new material in fact spurred him to develop inventive solutions and new types of imagery. As he explained,

> You do end up with certain kinds of mannerisms. You handle paints a certain way, you flick it this way and that way, and often this steals in unawares—without any deep implications; it just happens. It's just a flick of the wrist and it looks all right and you pass it and it stays there, and ultimately it spoils the painting. So I like to change the medium every now and again so that I can work against it; so that I am not proficient in it—because, in some way, I'm always worried by proficiency. (Nolan in Barber 1964, quoted in Klepac 2007)

Nolan was somewhat evasive about his painting methods and generally avoided inviting visitors into his studio. His second wife, Cynthia Nolan, wrote of a conversation she had with the Romanian American artist Saul Steinberg (1914–1999), whom she and Sidney met in New York in 1960:

> "Every word that Sidney says is camouflage [Steinberg remarked]. Everything he says goes off on a trail away from what he really believes, thinks, intends to do, has done in the past. He is like a bird dragging a wing, leading the hunter away from its young and nest."
>
> "Yes, that's true," Cynthia responded, "except for the times when he takes the pursuers and shows them the nest and even how he has constructed it."
>
> "Oh, but," Saul persisted, "that only fools them further. When he does that he has already abandoned that nest and built again, in another thicket. They will find only eggshells and the young birds flown." (C. Nolan 1967, 100)

Despite his apparent reluctance to share his technical secrets, Nolan's correspondence and interviews contain many clues to his methods and materials. Often these technical aspects have been overlooked in deference to an emphasis on the artist's conceptual vision and his personal life. Moreover, it is difficult to understand the complex terminology of modern paints, pigments, and trade names. Recent research by Crook and Learner (2000) and Standeven (2011, 2013) on the development of twentieth-century household paints has assisted in recovering some of this forgotten nomenclature. This work, along with the increasing ability of conservation laboratories to analytically identify paint media, has begun to unravel the mysteries of Nolan's practice. Nolan, himself intensely interested in science, would no doubt have enjoyed these advanced technologies being used to reveal some of his secrets. What emerges from this study is the extremely diverse range of materials and techniques that Nolan used in his long and prolific career and the incredible energy he had for new subjects and new materials with which to paint them.

1 St. Kilda Reflections: 1938–1941

Sidney Nolan grew up in a Melbourne bayside suburb called St. Kilda, a popular weekend destination with a Luna Park fairground, sea baths, and a long promenading pier. As a young boy, Nolan embraced these rich entertainment possibilities, holding particular affection for Luna Park (fig. 1). Circled by two wooden rollercoasters, the Scenic Railway and the Big Dipper, Luna Park had a distinctive perimeter of uprights, crossbeams, and curved tops that find visual clues in his paintings of the late 1930s and early 1940s (see figs. 4–6, 11). Memories of his St. Kilda youth resound in Nolan's work throughout his life. In 1977 Nolan stated, "I remember as a kid going to sleep feeling the ratchet sound of the Big Dipper going 'click, click, click' and the screams of the girls and I'd go to sleep and think, well, I just wish I was up there with them..." (quoted in Lindsay 2017, 68).

Despite his childhood and youth coinciding with the financial privations of the interwar years, Nolan held a lifetime affection for the places and people of his childhood, referencing them in his art throughout his career, and never ceased to express a

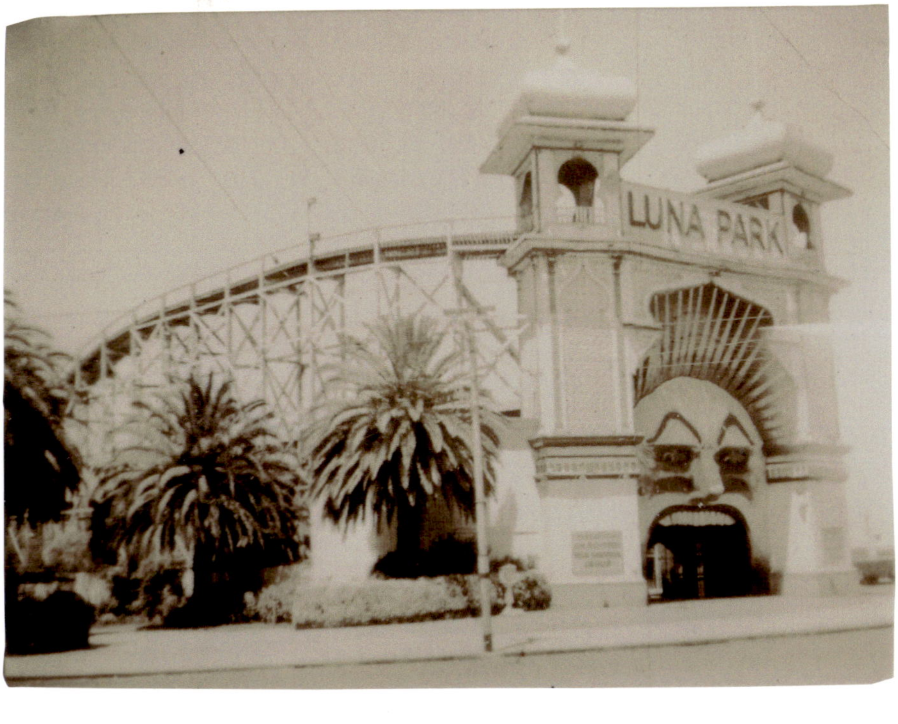

FIGURE 1
Albert Tucker, Luna Park, St. Kilda, c. 1939–45. Albert Tucker Photographic Collection, Heide Museum of Modern Art and State Library of Victoria.

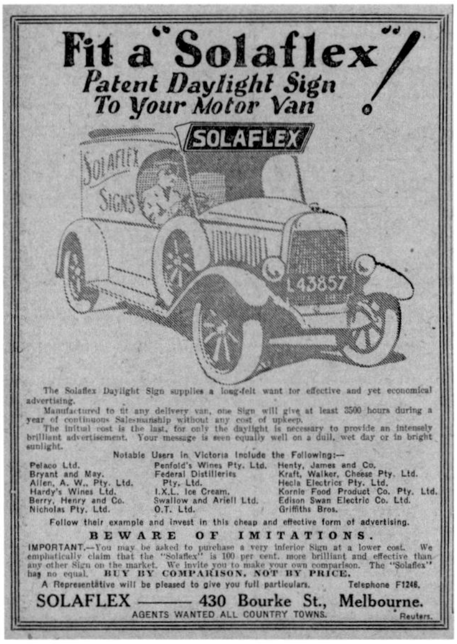

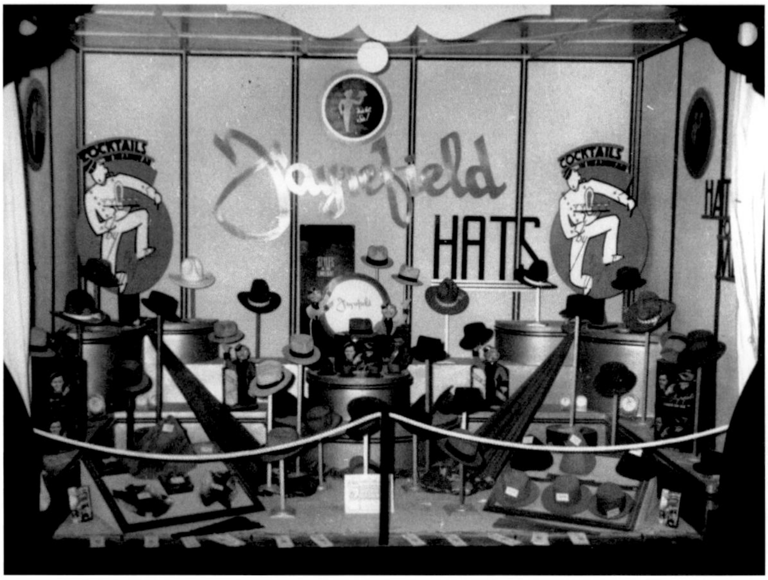

FIGURE 2
Advertisement for Solaflex, *The Argus*, February 10, 1928, 16. Newspaper Collection, State Library of Victoria.

FIGURE 3
Photographer unknown, Display of men's hats, Brisbane, c. 1937. Digitized negative from Red Cross Week window displays 1937–1944. John Oxley Library, State Library of Queensland.

sense of exuberant joy, a wry humor, and an irreverence for authority. He approached his education and working life determined to pursue his parallel interests in art making, writing, and cycling. In 1936 he rode his bicycle from Melbourne to Sydney, a distance of over a thousand kilometers (600 miles), pausing only briefly in Sydney before making the return trip.

In 1931, at fourteen years of age, Nolan found employment in the thriving area of advertising, thus fulfilling, at least initially, his need to make art: "My art set-up, you see, was in a way satisfied by the commercial art aspect of it," he remarked in a 1962 interview (B. Smith 1962). His first job was with a company called Solaflex, which made commercial signs for cars (fig. 2). His work involved cutting letters out of lead foil, which were used to mask glass panels; transparent paint was sprayed around the stencils onto the glass, which was then fitted into mirror-backed boxes attached to the tops of commercial vehicles. Sunlight shone through the lettering and was reflected forward by the mirror, creating an "illuminated" effect. Later in his career, Nolan transformed masking and working with reversed images into inventive approaches for art-making. After the collapse of the Solaflex Company in 1932, Nolan purloined some of the transparent paints he had been using, later using them in his own works (B. Smith 1962).

In 1933, Nolan started in the large art and advertising department of the Fayrefield Hat Company, gaining a more thorough grounding in commercial art practice. According to Vernon Jones, the Fayrefield Hats advertising manager, Nolan's job was to "cut out heads of cardboard, scale up lettering and paint it on window display cards" (quoted in Underhill 2015, 38). A photograph from about 1937 of a display of Fayrefield hats may show some of his work (fig. 3). Its cut-out shapes and metallic paint are strongly evocative of Nolan's later interests in masks, glossy surfaces, and spray paints.

Artist's Oil, Boot Polish, and Blotting Paper

In the late 1930s, Nolan had an epiphany, sparked by reproductions of works by Vincent van Gogh (1853–1890) and Pablo Picasso (1881–1973) that he saw in art books and magazines. Suddenly, he was attacked by a passion for abstraction and the desire to become an artist.

> Almost everything happens in the same year I think. About '37. And I became intolerable about it. Looking back on it, you know, I shoved it down everybody's throat. I would...I was sent...I got the message. And it was quite multiple. I mean, the distance between van Gogh and Picasso and everything else all came in six months...the whole changeover you know...I can't describe it, but I became obsessive about it. (B. Smith 1962)

Leaving his paid employment at Fayrefield Hats, Nolan embarked on the uncertain career of an artist. In 1938 he married a fellow artist, Elizabeth Paterson, who was trained in traditional fine-art painting at the National Gallery School in Melbourne. The two shared a studio at their home in a beachside town called Ocean Grove outside of Melbourne from 1938 to 1939. It is probable that they also shared painting materials.

A group of five paintings from Ocean Grove, including *Untitled (Abstract)* from 1939 (figs. 4–6), are painted on thick blotting paper. The papers all have a vertical crease down their center that was present before Nolan, undeterred by their imperfections, started to paint, employing a limited color range of black, blue, and red to reference the Luna Park rollercoasters of his youth. The ready absorbency of the blotting paper would have assisted in the rapid penetration of the solvent-thinned paint. In some places the color has bled all the way through to the back of the thick sheet, but this did not dissuade Nolan from turning the blotter over and beginning a new composition, sometimes deliberately singling out the bleed-through color as a focus of the design (see fig. 5). Occasionally, too, he superimposed a third image after applying a thin slurry of gesso chalk wash, leaving the first painting still faintly visible. Broad black borders on several of the works suggest that Nolan intended them to be lit from behind, perhaps to be held up against a window. In these cases, the three versions of the painting intersect to create yet another, a fourth, version of the work (see fig. 6). The masking provided by the borders and the multiple and reversed underlayers seen by illuminating the work from behind are strongly reminiscent of the techniques employed in the glass car signs Nolan had made for the Solaflex company.

The pigments analytically identified on the Ocean Grove blotter paintings are bone black (calcium phosphate), dinitroanaline red (Pigment Orange 5) mixed with chalk (calcium carbonate), and French ultramarine blue[1] (see sidebar on p. 16, FTIR and XRF Analysis, for a description of methodology). The brilliant orange/red dinitroanaline is a synthetic dye, one of a number of increasingly intense colors available through the twentieth century, synthesized from coal and petroleum products (see sidebar on p. 18: Coal-Tar Colors). While dinitroaniline is similar in hue to the more traditional red pigment vermilion (mercuric sulphide), its greater intensity is revealed when it is thinned to a translucent wash (see fig. 4).

While synthetic dyes such as dinitroaniline might be found in industrial and household decorative paints, bone black and French ultramarine pigments are less valued in those forms owing to their higher cost and poor tinting strength. Their presence in Nolan's blotter paintings suggests the use of artist's paints. The absence of significant FTIR (Fourier-transform infrared spectroscopy; see sidebar on p. 16) absorbance peaks for the binder[2] on analyzed samples and the lack of paint filler pigments also support the premise that these early paintings are made with good-quality artist's oil paints.

A painting by Nolan titled *Head of Rimbaud* (1938–39; fig. 7) has a more unusual medium description: "oil and boot polish on cardboard." The reference to boot polish suggests that the medium might be encaustic, that is, wax mixed with pigment or oil paint. However, the three colors of paint on the work, yellow, brown, and white, are all extremely hard, without the expected softness of a wax medium. Instead, the thick paint

FIGURE 4
Sidney Nolan, *Untitled (Abstract)*, 1939. Oil and gesso on blotting paper, 44.5 × 57.0 cm (17½ × 22½ in.). Art Gallery of NSW. Purchased 2013.

FIGURE 5
Verso of Sidney Nolan, *Untitled (Abstract)* (see fig. 4).

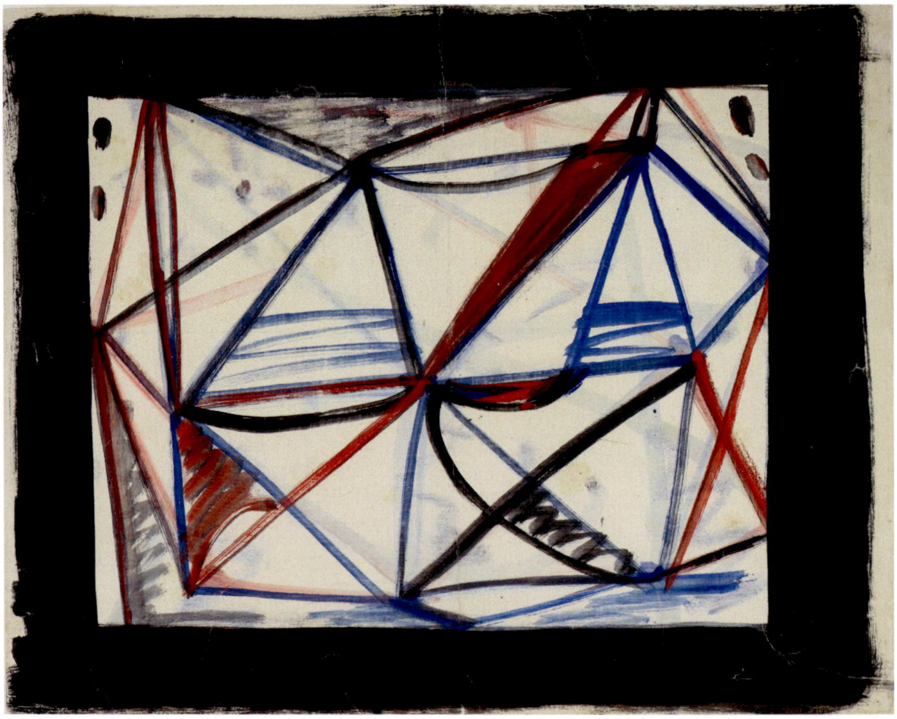

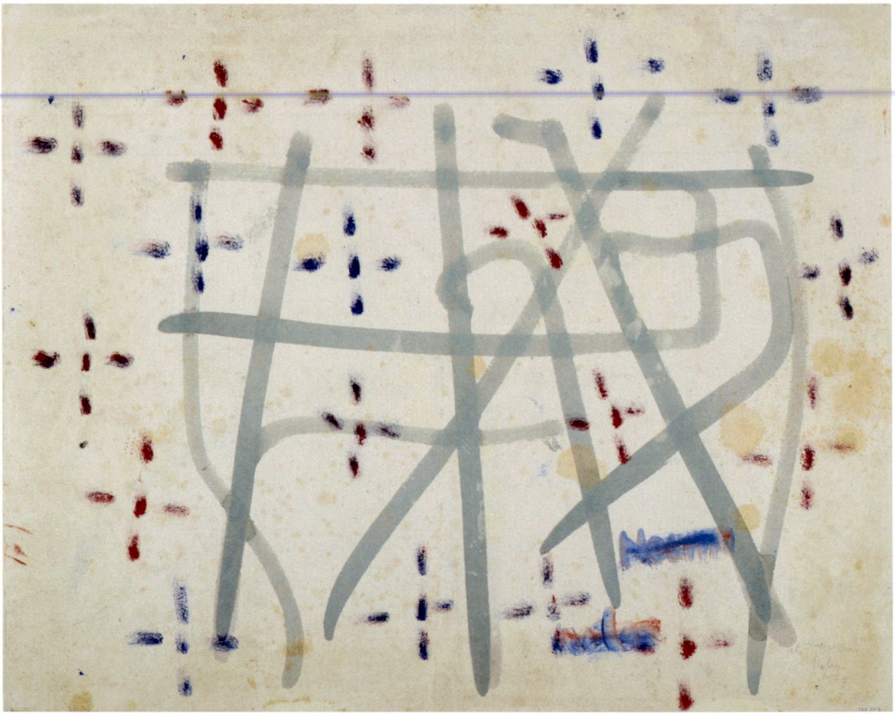

has held the sharp textural pattern of its application, the yellow and brown paints having been applied with a palette knife and the white stiffly brushed.

Pigment and binder analysis of paint scrapings on the back of *Head of Rimbaud* (the paint on the image itself being too intact to microsample for analysis) gave little information regarding the binder. The dark yellow paint pigments were identified as natural

FIGURE 6
Sidney Nolan, *Untitled (Abstract)* (see fig. 4) photographed on lightbox.

FIGURE 7
Sidney Nolan, *Head of Rimbaud*, 1938–39. Oil, boot polish, and pencil on cardboard, 34.3 × 26.9 cm (13½ × 10⅝ in.). Heide Museum of Modern Art. Purchased from John and Sunday Reed 1980.

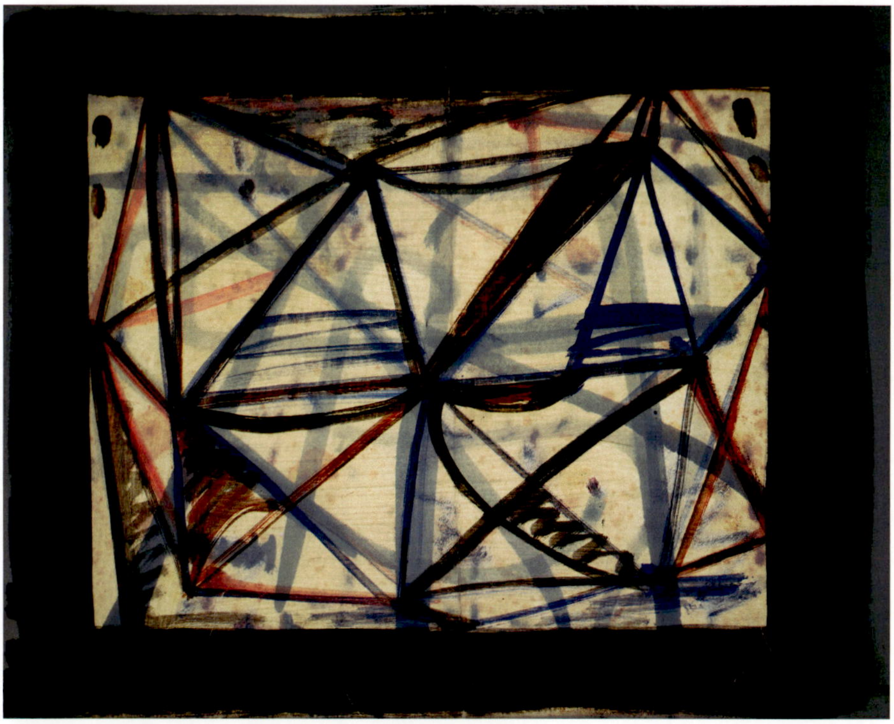

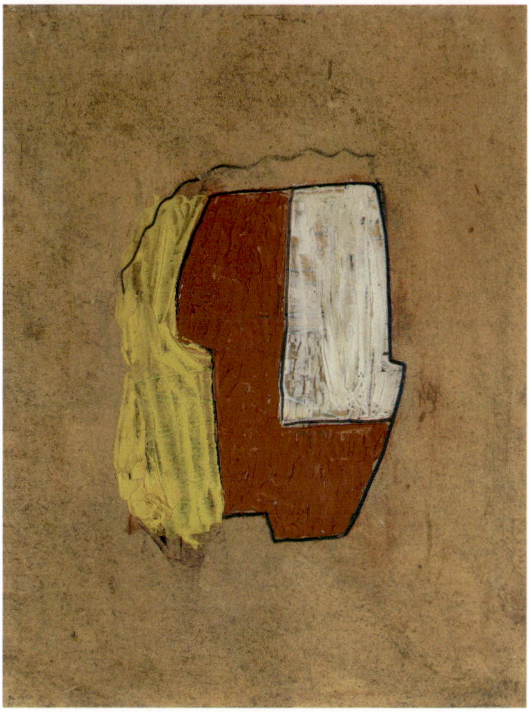

yellow ochre mixed with kaolinite (aluminum silicate), and the pale yellow as chrome yellow (lead chromate) mixed with gypsum (calcium sulphate dihydrate). The colored pigment in the brown paint could not be identified, but the filler pigments kaolinite, barites (barium sulphate), and chalk (calcium carbonate) were detected by FTIR analysis (see appendix I). The pigments' strong FTIR analytical response, the appearance and hardness of the thick pastelike paint, and the opacity of the colors are all features typical of artist's oil paints in which the pigments are present in high proportion to the binder. The presence of filler pigments suggests that the paints are different from the those used on the Ocean Grove blotter paintings, and of poorer quality, such as student-grade paints.

Close visual examination of the work suggests that the shoe polish may not have been mixed with the oil paint, but rather rubbed across the surface after some of the pencil design had been drawn on the cardboard, perhaps to seal and tone the support and drawing layer prior to painting. The use of the wax as a sealant is consistent with observations made by Nolan's biographer, Brian Adams (1987, 42), who noted that Nolan prepared the cardboard of the work with brown shoe polish prior to painting. However, if the wax layer was colored brown when Adams saw it in the 1980s, it has since faded.

At this time, Nolan saw the inevitable deterioration of his artworks as a kind of fey liberation.

> I was using boot blackening and everything under the sun because I believed at that stage that the more perishable the materials or [the] more unstable, the better. So there are practically months and months of work on tissue paper, with the deliberate intention that, because I liked doing it, but also the perverse kind of thing that ultimately it must perish, you know, for sure; that kind of attitude. (B. Smith 1962)

Set Design

In 1939 Nolan received a commission by the Original Ballet Russe company to design the set and costumes for a new one-act ballet, *Icare,* to be performed during the company's 1940 Australian tour. Nolan's set, unveiled at the show's premiere in Sydney, comprised a backdrop cloth in two parts, the one being stage forward of the other, allowing the dancers to move between the two flats (fig. 8). The final design was a progression of ideas documented through a series of works that may have begun with an unusual composite painting, *(Untitled-Abstract)* (c. 1939; fig. 9).

(Untitled-Abstract) is painted over a silver gelatin studio photograph of a woman in a Victorian-era dress holding a book on her lap. The sitter's left eye, which Nolan has left unpainted, is central to the composition and is referenced strongly in all the subsequent studies for the *Icare* set, including the final design. The found photograph acts both as support for a painted layer and as the image proper. This structural device of bringing different layers and compositions of a work into a single picture plane was a surrealist technique that Nolan used throughout his career. The paint applied over the photograph of *(Untitled-Abstract)* is a hard and thick paste type that has retained the texture of its brushed application. The pigments present are identical to those identified on the blotting-paper paintings, including the brilliant dinitroanaline red (PO5), suggesting that they may in fact be the same paints.

Although he was already experimenting with reused supports and unusual layering systems, these early works by Nolan appear to be rendered in the traditional artist's medium of oil paint. The use of recycled supports and boot polish are less typical and give clues to what was to emerge strongly in Nolan's work from 1940 on, a free approach to materials pushed technically in new ways, along with an interest in manipulating the gloss and matte effects of the paints by dilutions and polishes.

FIGURE 8
Hugh P. Hall, Roman Jasinsky as Icare, in *Icare*, Original Ballet Russe, Australian tour, His Majesty's Theatre, Melbourne, May 1940. Silver gelatin print, 19.5 × 24.8 cm (7¾ × 9¾ in.). Hugh P. Hall collection of photographs, 1938–40, National Library of Australia.

FIGURE 9
Sidney Nolan, *(Untitled-Abstract)*, c. 1939. Oil on silver gelatin photograph, 28.0 × 34.4 cm (11 × 13½ in.). Art Gallery of NSW. Bequest of Gwen Frolich 2006.

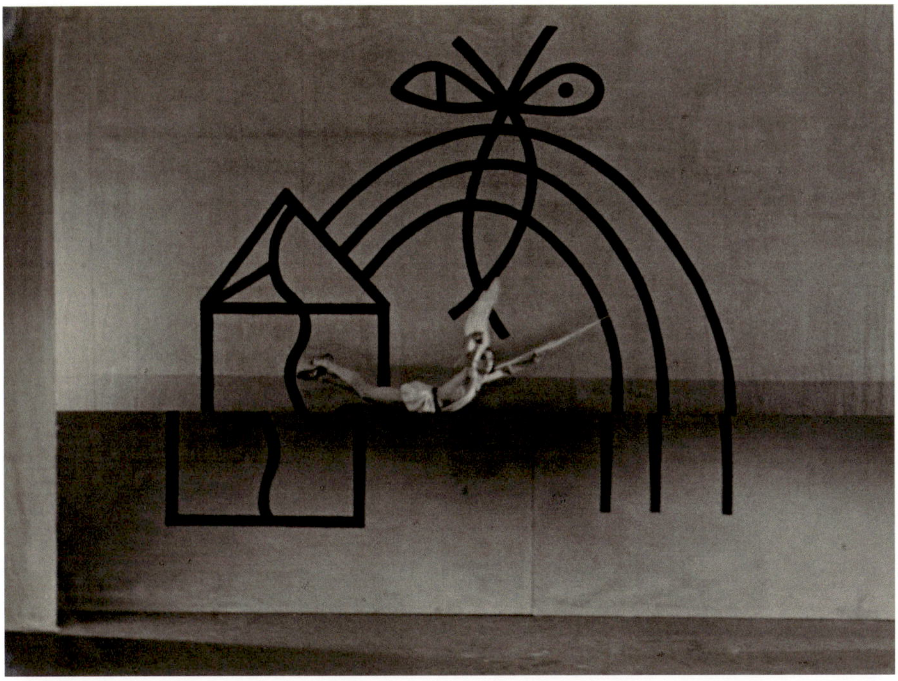

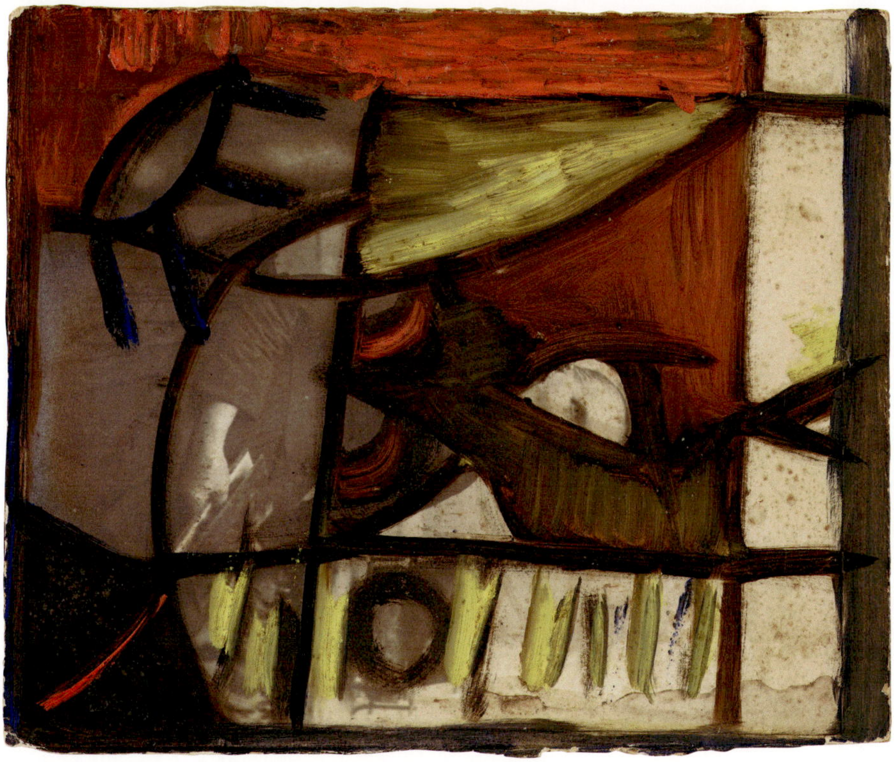

There is an exuberant pleasure in these works, in which Nolan found inspiration in everyday and discarded things. Resourcefulness, such as using old photographs, creased blotter papers, and cardboard for painting supports, may have been driven by economic necessity—having given up full-time paid employment, Nolan had little money to purchase artist's materials—but it was also driven by an irreverence for the traditions of painting and values of longevity.

Gloss Paint

The Herald Exhibition of French and British Contemporary Art, shown in Adelaide, Melbourne, and Sydney over the course of 1939, is said to have exposed many Australian artists for the first time to the work of such modern European artists as Picasso, Georges Braque (1882–1963), Paul Cézanne (1839–1906), and Henri Matisse (1869–1954). The exhibition opened at the state-run Art Gallery of South Australia in Adelaide (August 21–September 17, 1939) and—eschewing the National Gallery of Victoria and National Art Gallery of New South Wales—appeared at the Melbourne Town Hall (October 16–November 1, 1939) and the David Jones (department store) Gallery in Sydney (November 20–December 16, 1939). The exhibition attracted huge public interest, but its success was tempered by the outbreak of war, which created difficulty in returning the artworks to their owners at the end of the tour. As a result, they were kept in Australia for the duration of the conflict (Chanin and Miller 2005). Hung on the walls of various state galleries during their enforced stay in Australia, these works inspired a generation of Australian artists. Nolan visited the exhibition in Melbourne and spoke later of seeing the works of Salvador Dalí (1904–1989), Graham Sutherland (1903–1980), and Émile Bonnard (1868–1941), though he appeared to barely recall those by Picasso (B. Smith 1962). Yet it is difficult to imagine that seeing the paintings of Picasso in life, with their gloss paints and reduced surface texture—features not readily visible in reproduction—could have failed to have had a powerful influence on him.

After war was declared in September 1939, Nolan and Elizabeth moved back to Melbourne. Working nights in a hamburger grill, Nolan continued to paint but struggled to find financial and critical support. In 1940 he held his first exhibition, in his studio above a greengrocer's shop at 325 Russell Street, Melbourne. Nolan describes this exhibition of two hundred items as completely abstract, with "thousands of little bits of scribbly paper and everything all over the place" (B. Smith 1962). That year, 1940, also signaled a change in paint media as he moved away from the matte artist's oil paints to a new type of paint that was both liquid and glossy.

Artist's oil paint has specific properties unlike those of other types of paint. Generally, it is squeezed from a tube onto a flat palette and has a thick consistency that holds the textural pattern of the brush or palette knife with which it is applied to the canvas. Artist's paint is formulated for application to a vertical surface without dripping or running. Slow drying allows for blending and working wet into wet. The pigments and oils used in artist's paint are generally of the highest rating for light fastness and durability, and they are used in larger proportion to the binder, giving a high pigment loading. Because artist's paint is also used in much smaller quantities than household paint, it can make use of expensive pigments without a significant increase in cost. Until the late 1940s, artist's paints continued to be oil-based and supplied in metal tubes. Only in 1947 were plastic binders developed for paints manufactured specifically for artists, though these did not come to be generally used by artists until the 1960s.[3]

In the first half of the twentieth century, paint intended for house decorating, advertising signage, and motor vehicles saw rapid technological development. Previously, most household paints had been mixed by hand by professional painters from pastes and seed oils and tinted with pigments. As the need for paint to perform in specific ways was extended, the industry developed many different types of paints ready-prepared for the user. Beginning in the 1920s new paint binders were also chemically synthesized, many of which had specific advantages over traditional oil paints. The two most successful of these, nitrocellulose (see below) and alkyd (see also chapter 2), were gloss paint media initially developed for spray-painting cars. The attraction of these fast-drying gloss paints ensured that they were quickly reformulated for use as brushing paints for building, decorating, and commercial art (Standeven 2011).

While some artists have been known to use industrial and household paints on their artworks, until recently their use of these materials has usually been confirmed by visual characterization alone, or sometimes with the corroboration of an artist's records. Features such as a lack of stiff paint texture or the self-leveling of liquid paints into flat surfaces have been used to suggest the presence of non-artist's paint. This methodology is problematic, since, when dry, glossy enamel paints made from oil tend to be indistinguishable in appearance from those made from alkyd and nitrocellulose. It has also been demonstrated that paint features associated with the use of household decorative paints, such as self-leveling, inherent gloss, and liquid application, can all be replicated by the addition of solvents, oils, and resins to standard artist's oil paint (Carlyle 2011). Fortunately, micro-FTIR can distinguish these different paint binders effectively using microscopic-sized samples taken carefully from paintings at edges or near preexisting damage. (See sidebar on p. 16: FTIR and XRF Analysis.)

Household paints are fast-drying and liquid. They need to be fluid to allow spreading and ensure an even finish and coverage. Quick drying after application is important to avoid dirt pickup and to enable recoating. While durability is a concern, it is tempered by the need to keep the cost low; therefore, cheap, high-tinting pigments are used in preference to the more expensive and permanent pigments found in artist's paint. In addition, because the larger volume required of commercial paint makes it more vulnerable to raw material shortages and the shifting politics of international trade, the formulation is more likely to change over time. Despite the ready availability of household and commercial paint compared to artist's paint, therefore, their different application and physical properties resulted in limited adoption by most artists. Nolan, however, had already worked with these paints as a commercial artist and saw their potential in his new career as a fine artist.

In 1941, Nolan began work on a large group of paintings on slate tiles that he removed surreptitiously from the roof of the studio he was renting in Melbourne. Starting in January and continuing into 1942, he painted thirty-two tiles—on some occasions using both sides—with five colors: brilliant white, rich crimson, bright mid-blue, rich yellow, and black. The paint shows a variation in gloss levels across the group, probably due to the differences in the slates' ability to absorb the paint, but in general the surfaces are glossy, suggesting the paint was fluid when applied, and are without the thickness and texture expected with the use of traditional artist's oil paint. Analysis of a tiny fragment from an untitled painting dated December 11, 1941 (fig. 10) gave a positive result for the presence of nitrocellulose. While it is possible that Nolan used different paint types as he worked on these paintings over the year, the distinctive palette of limited colors, particularly the deep crimson, is used throughout, making it likely that many, if not all, of the slates were painted using this same medium.

Samples of the paint on another work from 1941, a painting on canvas called *Luna Park* (fig. 11), were analyzed more thoroughly. Three colors gave the distinctive result for nitrocellulose lacquer. It is difficult to understand how Nolan could have worked with such a fast-drying paint, which he brushed and mixed directly on the canvas. Unfortunately, the original gloss of the paint is no longer visible under varnish layers applied during later conservation treatments. The white pigment in nitrocellulose

FIGURE 10
Verso of Sidney Nolan, *Untitled (Figure and Bird)*, December 11, 1941. Nitrocellulose lacquer on slate, 25.7 × 50.8 cm (10¼ × 20 in.). University of Queensland Art Museum. Purchased with the assistance of the Alumni Association and the Peter Stuyvesant Cultural Foundation, 1977.

FIGURE 11
Sidney Nolan, *Luna Park*, 1941. Nitrocellulose lacquer on canvas, 67.0 × 84.0 cm (26⅜ × 33⅛ in.). Art Gallery of NSW. Purchased with funds provided by the Nelson Meers Foundation 2003.

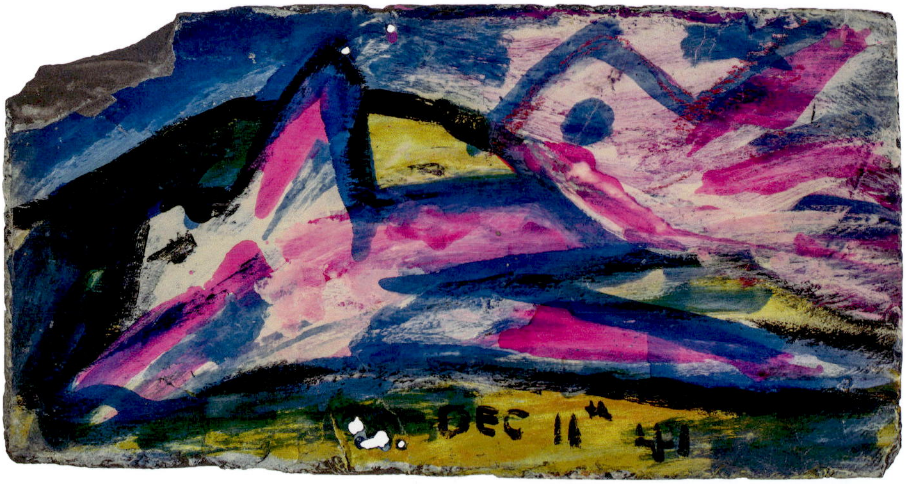

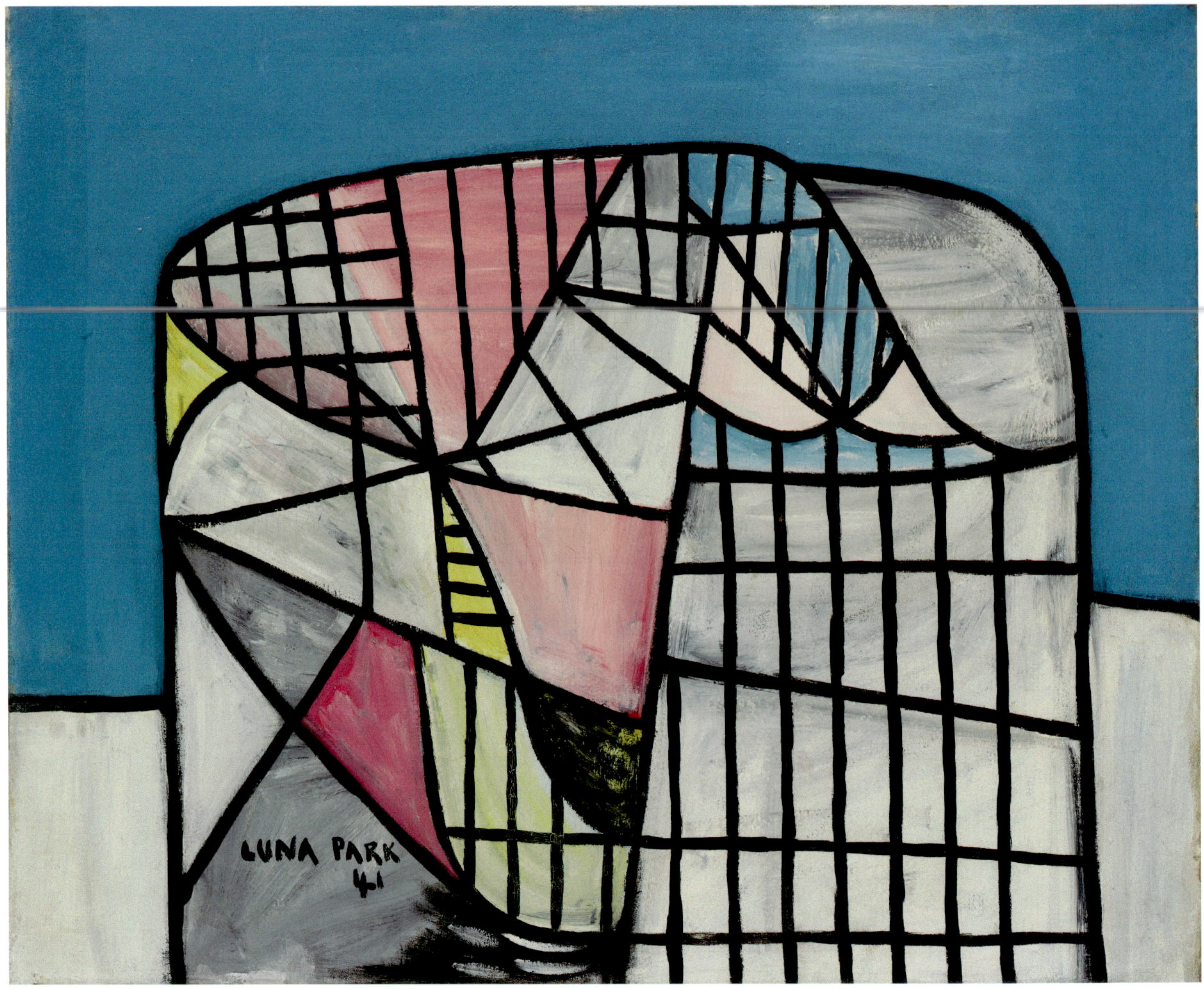

FIGURE 12
Sidney Nolan, *Untitled (Round Tree)*, c. 1941. Nitrocellulose lacquer on plywood, 38.1 × 25.4 cm (15 × 10 in.). Art Gallery of NSW. Purchased 2013.

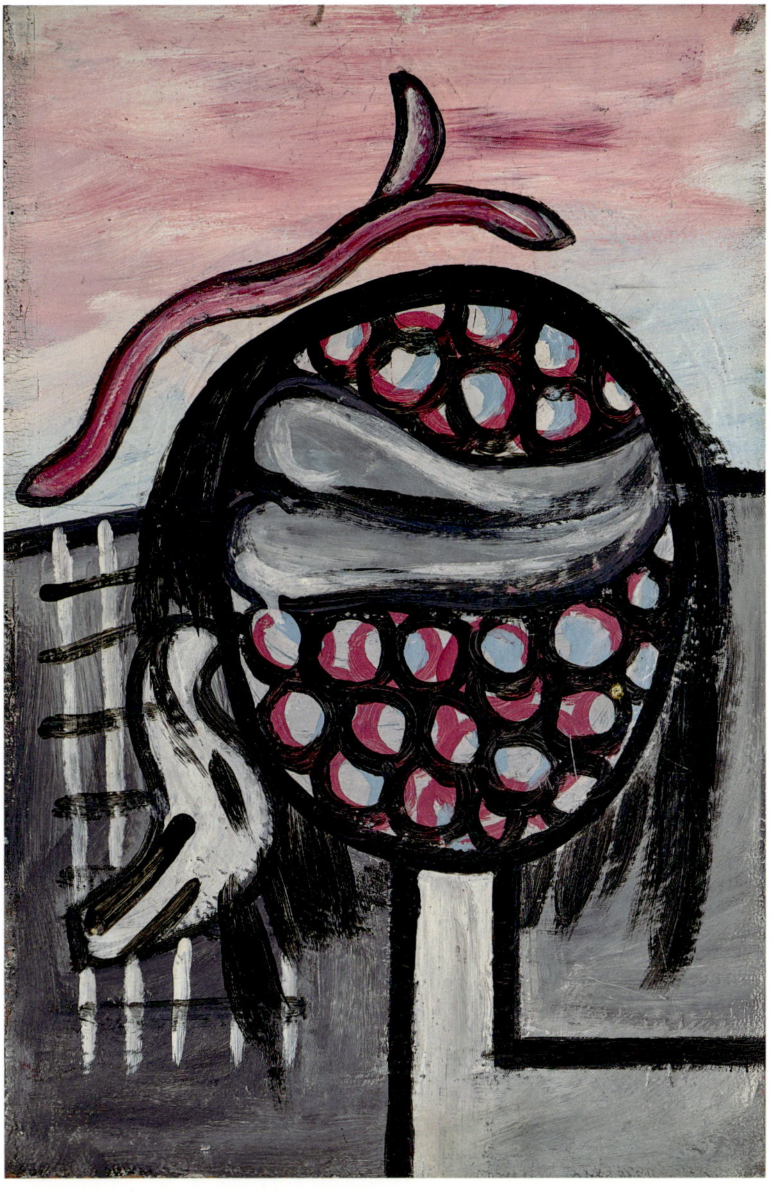

paints is typically titanium white (titanium dioxide), appreciated for its opacity and fine particle size; this pigment renders *Luna Park* brilliant in its cool tonalities. Two further 1941 paintings by Nolan with similar color palettes have also been tested and identified as nitrocellulose paintings—*Bird* (December 4, 1941) and *Untitled (Round Tree)* (c. 1941; fig. 12)—suggesting that nitrocellulose lacquer was Nolan's principal paint medium through 1941.

Nitrocellulose Lacquer

Elaborate window displays were an important part of the modern cityscape in the 1930s. Advertising had become a huge industry, and brightly colored and glossy paints played an important role in inciting desire through the suggestion of glamour, luxury, and fashion. The paints used in such displays were quick-drying spray paints based on

FTIR and XRF Analysis

FTIR is a well-established technique for the characterization of paint binders, having been used in the commercial paint industry for decades. More recently, FTIR microscopic instruments, in which extremely small samples can be analyzed,[4] have become commonplace in conservation laboratories as essential tools for identifying the different types of paint binders found in art works.

Oil, alkyd (oil-modified polyester resin), and nitrocellulose each have distinctive spectral absorptions in the infrared region that can be assigned to the different chemical bonds in the compound (fig. 13). The arrangement and presence of a combination of bonds, seen as absorption peaks in an FTIR spectrum, can be matched to known reference samples (table 1). Some common pigments also demonstrate infrared absorption features in FTIR. These may be useful for identifying pigments but can obscure the absorption peaks associated with the binder in paint samples. Fortunately, most commercial paints, especially glossy ones, have a high proportion of binder and use smaller amounts of strong tinting pigments compared to artist's oil paints. This assists in the identification of the binder in commercial paints. An additional difficulty may arise in interpreting the FTIR results from paints on artworks, in that pigments may interact with the binder to form secondary products that can be difficult to assign directly to a pigment or binder. This is particularly the case with oil paints, which can have complex FTIR spectral results.

The presence of alkyd on artworks prior to 1963 or nitro-cellulose at any time indicates the presence of a non-artist's paint. The distinction of artist's oil paint from a commercial oil-enamel is less definitive. Enamel will typically have a higher oil content than an artist's paint, making the oil binder more readily identifiable in a pigmented sample.

X-ray fluorescence spectroscopy (XRF) is an analytical technique in which the surface of an artwork is exposed to a flow of X-rays. As these high-energy wavelengths engage with the pigments in the painting, the elements present are excited for a microsecond. The release of this energy during a return of the elements back to their non-excited state is detected by the instrument, and thus the elements present in pigments can be identified by the wavelengths of the released energy. This is a useful technique for identifying inorganic metallic pigments that may not be responsive to infrared absorbance.

All analytical techniques have their specific capabilities in identifying pigments and binders in complex paint mixtures. Ideally, each paint color on an artwork would be tested with several different techniques to ensure that the full range of materials are identified and verified.

new types of resins and dyes. The most popular paint type was nitrocellulose lacquer. Developed initially as a strengthening and sealing coating for cloth-covered airplanes, it become a popular colored paint in the 1930s.

Nitrocellulose is a semisynthetic, derived from natural cellulose fiber that undergoes synthesis with nitrate. Paints made from nitrocellulose were called lacquers, as drying and film formation occurred by solvent evaporation alone, not slow oxidation and polymerization as with traditional oil-based paints. Available from the early 1920s, colored nitrocellulose paints revolutionized the car industry. Previously, cars were painted with oil-based enamels, which were slow-drying and required many coats for durability, or phenolic resins, which could only be made in dark colors. Nitrocellulose paint dramatically reduced production time, allowing the painting of the metal bodies to be completed in days rather than weeks, and the lacquer could be colored with a wide range of pigments, giving rise to brightly colored cars for the first time.

DuPont had marketed the first nitrocellulose lacquer paint, Duco®, in the United States in 1923 (Standeven 2011). Available shortly afterward in Australia, Duco as a finish for cars was manufactured from October 1928 at the British Australian Lead Manufacturers (BALM) factory in Sydney (Todd 1998). Other Australian paint manufacturers

FIGURE 13
Typical FTIR absorbance spectra of (a) nitrocellulose lacquer, (b) alkyd, and (c) drying oil.

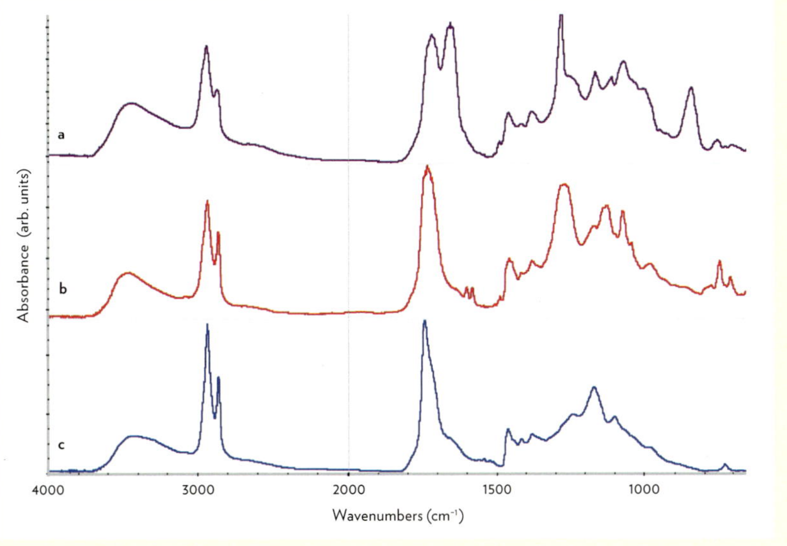

TABLE 1 Typical FTIR absorption peaks (wavenumbers cm^{-1}) for three common types of industrial and household gloss paints available in the 1930s (distinctive and strong absorption peaks indicated in bold)

Nitrocellulose	2932, 2860, **1717**, **1654**, 1458, 1378, **1279**, 1165, 1110, 1067, 842, 754
Alkyd	2928, 2855, 1731, **1599**, **1580**, 1456, 1378, **1268**, 1123, 1071, 744, 706
Drying oil	2927, 2855, 1740, 1461, 1377, **1168**, **1099**, **725**

soon began to produce their own pigmented nitrocellulose paints. The strongest competitor to Duco was Opex™, manufactured by Berger Australia through licensing arrangements with the American company Sherwin-Williams. This new paint boasted, in addition to high gloss and fast-drying characteristics, an intense and extended range of colors. The use of colored pigments with nitrocellulose resulted from the increasing number of different chemical derivatives from coal tar (see sidebar on p. 18) and the availability of the new white pigment titanium dioxide that provided opacity to the colored pigments and was fine enough to pass through the nozzle of the spray gun applicator without causing clogging.

However, the availability of cars with brightly colored high-gloss finishes created a new type of problem as customers found themselves overwhelmed by too much choice. When the Ford Model A was launched in Australia in 1928 using Berger's new Opex nitrocellulose paint, three Australian artists were enlisted to create color charts of possible options: George Lambert, representing an older-style English gentleman; Sidney Ure Smith, the city man; and Thea Procter, the modern woman. Consumers could simply specify a prescribed color arrangement, thereby avoiding both the agony of decision-making and possible regret at the outcome of an arbitrary color choice.

Coal-Tar Colors

Although many traditional pigments were used to color nitrocellulose paints, such as the cheap chrome yellow (lead chromate) and the strong-tinting dark blue pigment Prussian blue (iron ferrocyanine), bright new dyes derived from coal gave intense, brilliant colors, though they could vary in permanence. The identification of coal-tar colors is useful not only for dating paintings (by comparison with known dates of production of the various different compounds) and acknowledging an artist's adoption of new materials, but also in considering lightfastness and permanence.

Coal-tar colors were often given different brand names when they were manufactured by different companies, making it difficult to unravel the complex terminology of related chemical compounds. Hansa yellow 10G, for example, is manufactured by Hoechst in Germany; yet it is the same compound as Monolite™ yellow, manufactured by Imperial Chemical Company (ICI), and when used in paints manufactured by the artist's paint maker Winsor & Newton (UK), it is named Winsor yellow. It is therefore most accurately described by the Color Index Number, Pigment Yellow 3 (PY3), which refers to all of these brand-name colors made from monoazo arylamide.

Artists took an interest in the development of these brightly colored and shiny paints and their dramatic influence on the cityscape through car paint, signage, and architectural design through the late 1920s and 1930s. Roy de Maistre (1894–1968) and Hera Roberts (1892–1969), for example, used nitrocellulose lacquers to produce brightly colored furniture (McPhee 2013; Weise 2013). Other artists took nitrocellulose-painted furniture as subjects for paintings, as in Grace Cossington-Smith's *The Lacquer Room* (1936), depicting a tea room with shiny bright red tables and green chairs. Yet there is no evidence that nitrocellulose paints were used directly for making paintings in Australia prior to Nolan's slate works in 1941.

In the United States, however, the Mexican American artist David Alfaro Siqueiros (1896–1974) was gaining a reputation as an advocate for the artistic use of nitrocellulose lacquer. As an open critic of Mexico's governments through the first half of the twentieth century and a supporter of revolution, Siqueiros enjoyed the paint's pedigree, emerging as it did from the development of nitroglycerine explosives. He saw no separation between art practice and political conviction and welcomed nitrocellulose's nonartistic, commercial, and industrial roots, countering as they did traditional norms of art making (McGlinchey et al. 2013). In 1936–37 he ran a number of experimental workshops in New York to teach other artists, including a young Jackson Pollock (1912–1956), how to paint with this fast-drying liquid medium, pouring the lacquer directly onto supports laid flat or spraying it through stencils. In 1939 Siqueiros held a well-publicized New York exhibition of his spray-painted nitrocellulose paintings, which was reviewed in the January 1940 issue of *Art News*. This exhibition and its subsequent review may have influenced Nolan's own adoption of nitrocellulose paints for artworks in 1941.[5]

While longevity was not a great concern in Nolan's early experiments with non-artist's paint types, it became so over the next few years. Nolan's interest in transient and fugitive materials was changing, and he began to value more durable, but no less experimental, options. The change in attitude, as we shall see, was probably due to the influence of an important new person in Nolan's personal and professional life: Sunday Reed (1905–1981). Nevertheless, as the Second World War continued into 1942, and his circumstances grew more limited and supplies more difficult to procure, Nolan's preference for recycled supports and commercial gloss paint products became a versatile strategy.

2 Wartime Wimmera: 1942–1944

The Paint Manufacturer's Lament

If you formulate varnish and paint
You have excellent cause for complaint
When you think you are choosing
What you will be using
The Government tells you "You ain't!"

If there's Chinawood oil in your stuff,
To get it is terribly tough.
Dehydrated castor
They tell you is faster,
But still there is never enough.

When you pick out your resins with care,
Priorities get in your hair.
Your set-up they "bollix"
By cutting phenolics
And leaving you up in the air.

Of course there is nothing to do
But try formulations anew.
You cannot use vinyl—
That order is final—
You'll have to get some other "goo."

If lacquers you're trying to make,
You'll find it is all a mistake,
Solvents and cotton
Must all be forgotten,
Each number you buy is a "fake."

Of all the pigments we lack
No human could ever keep track.
The colours are many
And you may choose any—
So long as you're sticking to black.

The Army and Navy, 'tis true,
Want plenty of coatings from you,
But specifications
Perform such gyrations
You never know what you can do.

There's only one comforting note—
We're all in the very same boat,
So make up your lacquer
Of cheese and tobaccer
And don't let it all get your goat.

(Dudley Clapp, New England Paint and Varnish Club, quoted in Boland 1944–45)

Drafted

After leaving Fayrefield Hats in 1937, Nolan actively sought a patron for financial backing to establish himself in a new artistic career. He found it in John and Sunday Reed. The Reeds lived in a farmhouse on the suburban edge of Melbourne, at a place called Heidelberg. In the late nineteenth century the pastoral landscape of the area had inspired a group of landscape painters who came to be known as the Heidelberg school,

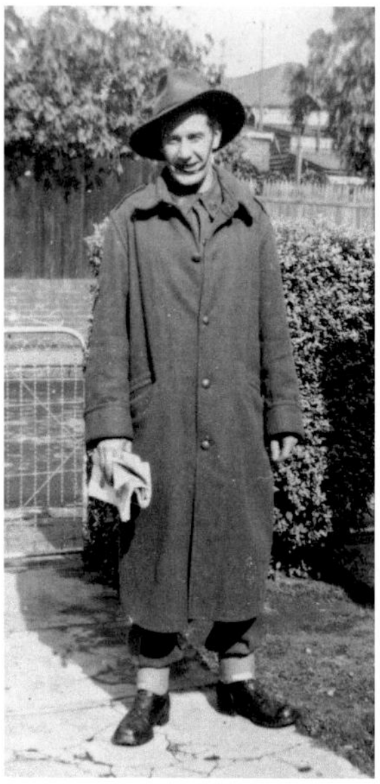

FIGURE 14
Photographer unknown, Private Nolan, c. 1942.

including Arthur Streeton (1867–1943), Tom Roberts (1856–1931), and Charles Conder (1868–1909). The development of a school of painting that took the Australian landscape as its subject and found new ways to depict its unique topography, color, and light was inspiration for the later generation of artists and writers who gathered around the Reeds at their home, which they named Heide.

Nolan's interests and focus began to drift toward this group of people centered around the Reeds, in particular the artist couple Albert Tucker (1914–1999) and Joy Hester (1920–1960). Tucker was an important mentor for Nolan, not least for his technical knowledge of painting and his willingness to try new and innovative paints and media. Nolan's marriage to Elizabeth broke down just before their daughter, Amelda, was born in January 1941, owing to a combination of pressures that included financial strain and Nolan's new relationship with the Reeds. Nolan soon moved into Heide, where he was supported both financially and creatively by the Reeds, who fostered his drive to make art. Nolan and Sunday also became lovers.

April 1942 brought another dramatic change in Nolan's circumstances when he was conscripted into the Australian Army (fig. 14). The war had already frustrated his plans of moving to Paris to pursue his artistic ambitions, and now he had to submit to the limitations imposed by army life. Nolan was sent for training in Seymour outside Melbourne. Deployed several weeks later, he was fortunate in being spared an overseas posting to the battlefields; instead he was placed as a private in the Supply and Personnel Column in western Victoria, in the area known as the Wimmera. Letters he wrote to Sunday Reed through 1942 state that he was part of a labor gang, hauling logs, peeling potatoes, painting trucks in camouflage, and undertaking guard duty.

An unexpected outcome of this imposed service was Nolan's embrace of the Wimmera, which inspired him to take up a new genre of painting: the Australian landscape. The Wimmera is a flat, semiarid region with few natural features to break the long horizon line. This topography was a revelation to Nolan, who began to reimagine the Australian landscape in a new modernist idiom. His task was evident from the moment he first settled at his base in the small town of Dimboola:

> Arrived here tonight at 6.30. I was wondering coming up in the truck how I was going to convey to you the flatness which I have never seen before and how one could do anything about it or live amongst it. It was alright while we were in sight of the Grampians and then suddenly was this feeling that left nothing of the earth except a thin line. And while I was thinking about all these things it came, simply that if you imagine the land going vertically into the sky it would work. (Nolan to Sunday Reed, May 4, 1942)

This compositional framework of narrow skies and high horizons became a scheme Nolan used in many of his paintings of the region, such as *Wimmera Landscape* (1943; fig. 15). It also had a sustained influence on the subsequent generation of Australian artists, in particular the painter Fred Williams (1927–1982), who used the same motif in his landscapes and was himself to play a critical role in the future conservation of Nolan's paintings (see chapter 3).

Throughout 1942 Nolan moved around among the small Wimmera towns where he was based, traveling between Horsham, Ballarat, Dimboola, and Nhill, with occasional weekends on leave spent back at Heide with John and Sunday Reed. Regular letters to Sunday are filled with thoughts of the heart and his longing to be with her, as well as descriptions of the country and of his explorations within it. He wrote extensively of walking in the evenings, the bird life, and the color of local places.

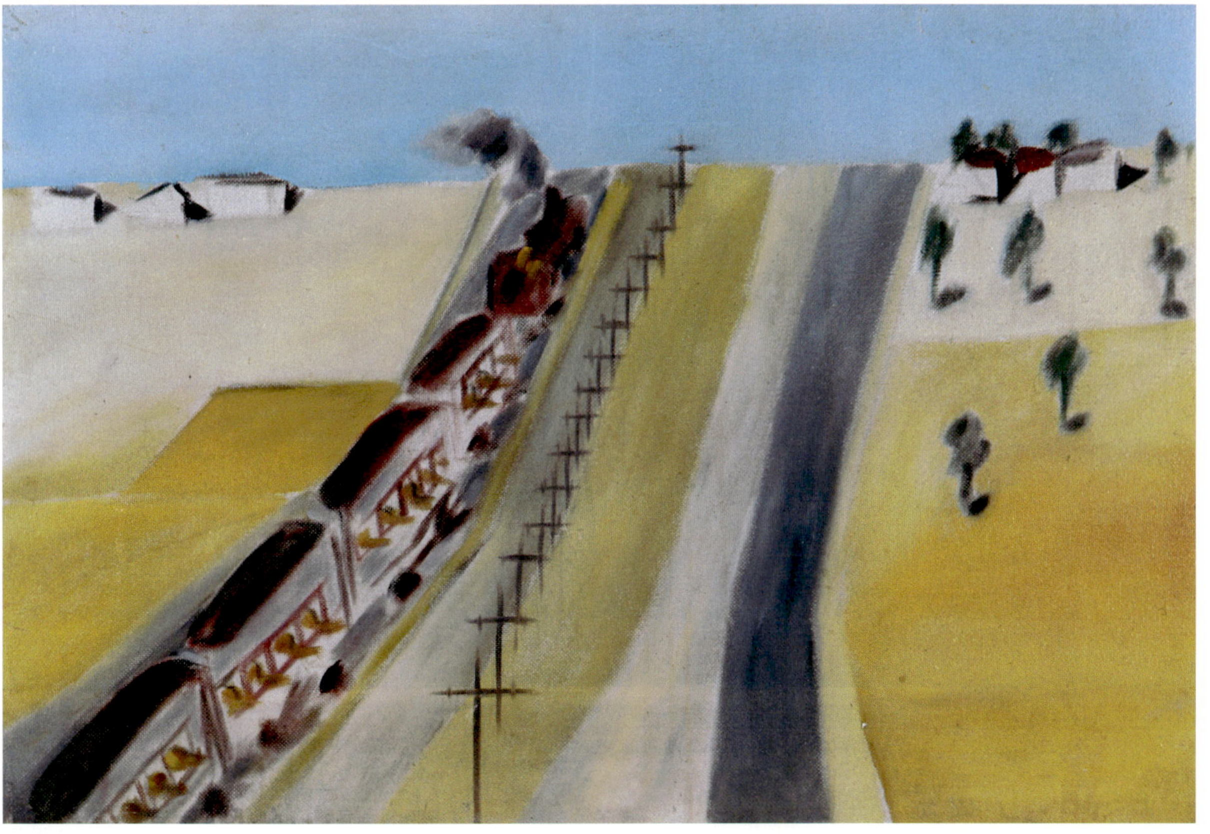

FIGURE 15
Sidney Nolan, *Wimmera Landscape*, 1943. Ripolin oil enamel on Masonite hardboard, 60.8 × 91.4 cm (24 × 36 in.). Heide Museum of Modern Art. Purchased from John and Sunday Reed 1980.

Nolan also wrote of his efforts to be reassigned as a war artist. In addition to giving him noncombatant status, this role would have enabled him to engage fully with painting and gain access to difficult-to-obtain art materials. Yet his efforts came to nothing. In August 1942, writing from Horsham, he expressed relief at being at least engaged to paint camouflage on trucks: "On the move again darling. End by being a gypsy painter. Down to Ballarat while the results of the report go through whatever they are. Camouflaging trucks I [thank] God to believe. Well I feel it's a start anyway & the break here had to come" (Nolan to Sunday Reed, August 1942).

Despite being unable to find a professional artist role within the army, Nolan was painting his own works during his period of service. In the first six months his efforts were principally in the portable mediums of watercolor, chalk, and ink on paper, but by August 1942 he made a shift toward solvent-based paints, as recorded in his letters and in Sunday Reed's diaries.

Diaries and Letters

An extraordinary record of Nolan's wartime efforts to continue to paint, and to find materials and new subjects, and of his collaboration with Sunday in these efforts, is recorded in three diaries maintained by Sunday in 1942 and 1943 and a collection of letters from Nolan to Sunday written from his various bases through 1942 to 1943.[1] The diary entries in Sunday's hand note the dates on which letters and telephone calls were received, Nolan's whereabouts and leave status, and critically, the dates that Sunday received Nolan's paintings for safekeeping and critical feedback.

FIGURE 16
Sidney Nolan, *Moonboy*, 1940. Alkyd on velvet on board, 55 × 57.5 cm (21⅝ × 22⅝ in.). Heide Museum of Modern Art. Bequest of Barrett Reid 2000.

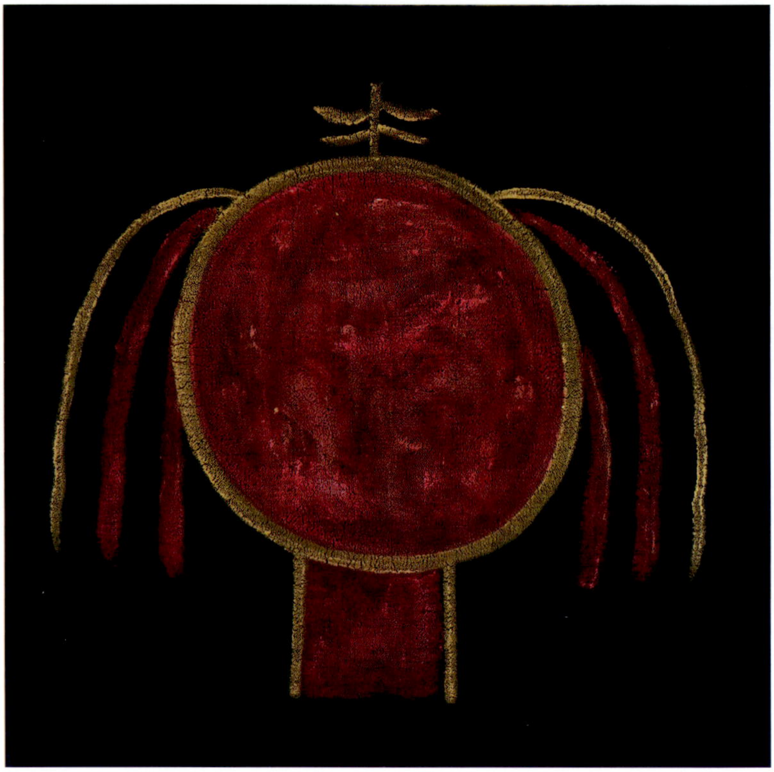

Beginning her first entry on April 15, 1942, with the declaration "Nolan went to war," by August Sunday includes a suggestion that Nolan was beginning to work with solvent-based paints, citing three small black-and-white paintings on tin that he had sent her. On September 8, Sunday writes that she "received lacquer and sketchbook," and on September 26, a painting in lacquer on the back of an envelope. While she notes receipt of several other small paintings through October, it isn't until the end of that month that she begins to consistently record the media and the titles of works in her diary.

October 1942 also corresponded with Nolan's promotion to corporal, as announced in Reed's diary on the twentieth and officially recorded on his service record on November 3.[2] Nolan's new, more settled lodgings, assisted by the advance in ranking, enabled him to start to paint in media that required a more permanent setup. As he remarked in an October letter to Sunday, "I wrote to Ron to ask Bilby [Nolan's commanding officer] to bring my paints with him. There is a room next to the office here empty & I can start all over again, travelling painter No. 1."

Dulux

Prior to his life in the army, in 1940, Nolan had used the household paint Dulux® on a work on black velvet cloth titled *Moonboy* (fig. 16). An inscription on the back in Nolan's hand records the paint brand (Harding and Morgan 2015). Dulux was a new type of glossy household paint that utilized alkyd as a paint binder. Alkyds, made from polyester plastic solubilized in oil, are the first true synthetic paint. They were available as a brushable household paint in Australia, the United States, and Europe from the early 1930s but have not previously been recorded as used by artists until the late 1940s.

Alkyds have been identified on paintings by Jackson Pollock (1912–1956) dating from 1949 (Schilling, Mazurek, and Learner 2007, 137) and Willem de Kooning (1904–1997) in 1948 (Lake 2010). Nolan's use of Dulux in 1940 is the earliest record known to the author of an artist employing a synthetic alkyd paint.

As he sought out paint in 1942, Dulux was again on his mind, and he wrote to Sunday asking her to send him some from Melbourne:

> The lacquer position is not bright but here are the main outlines to work on, apart from what you might discover in your travels. Dulux is probably the most durable lacquer or enamel on the market,[3] so is really the best angle to concentrate on. Their three best colors, strongest, brightest, are lemon yellow, cobalt blue & the red I've forgotten but you can tell it easily enough on the color card. Those three colors & black & white & thinner give a pretty complete range. (Nolan to Sunday Reed, October 27, 1942)

Looking at a prewar Dulux color chart made by British Australian Lead Manufacturers (BALM) (fig. 17), we might surmise that Nolan's request to Sunday indicated the Bright Red 5302, Cobalt Blue 334, and perhaps Pale Gold 012 (a lemon yellow is not present). The chart also illustrates the problem for artists of the limited color range offered in most household paints and their tendency to browns and dark tones, typically used for house trim, rather than the bright, high-key colors available in artist's paints.

A critical issue in the use of household paints by artists, particularly prior to the development of in-store tinting machines in the 1950s ("Paint Bar Where Colours Are Mixed" 1953, 22), was that the colors prepared by the manufacturers came in a limited range and each was individually formulated to achieve similar qualities of drying time, gloss, and thickness. The mixing of different colors together was not recommended as the different-colored paints, even those of the same brand, might not be compatible. This might have been a shortcoming for most artists, but Nolan worked with primary colors straight from the can, applying the paints with little premixing or modulation of tone on the canvas or board.

FIGURE 17
Prewar BALM Dulux 388 Brushing Line color chart. Caroline Simpson Library and Research Centre, Sydney Living Museums.

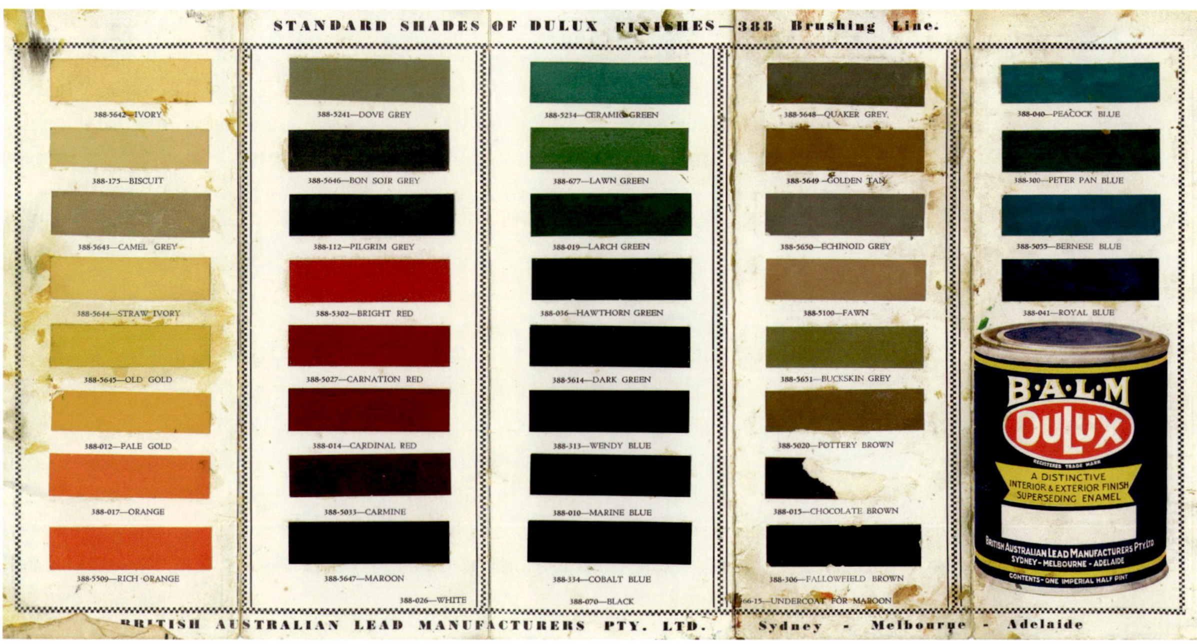

The illustration of the paint can on the color chart—"A distinctive interior and exterior finish," it reads, "superseding enamels"—also makes a critical point: Dulux was different. It was a synthetic paint, not an enamel or a lacquer despite its glossy self-leveling appearance. Developed by the American company DuPont in 1928 as an alternative to nitrocellulose, the more durable alkyd binders (also described as alkydals or glyptals, depending on their source) soon began to supersede nitrocellulose lacquers and oil enamels as a gloss paint for cars.

In 1929, BALM began production of Dulux at its Holden car factory in Cabarita, a suburb of Sydney. As the recession began to hit in 1930 and car sales declined, the product was diversified into a household paint called the Dulux 388 brushing line ("Dulux launch" 1931).

Other paint makers quickly caught on to the alkyd advantage. Taubmans began importing German-made alkyds in the early 1930s, soon formulating a new paint line named Dynamel™, and Berger produced an alkyd paint called Synthelac™. The paint industry's quick adoption of this revolutionary new synthetic paint was due to various factors: the harsh environmental conditions in Australia, which made durable alkyds a better alternative; a nontraditional house painting profession willing to try new materials; and a population of keen do-it-yourself home renovators for whom "synthetic" was an attractive concept. As a visiting paint chemist to Sydney in 1937 remarked,

> Australia has taken up synthetics as quickly as America, and quicker than England. I think you have even taken them up faster than America. In Australia, you have a sunny climate. There are about 300 days in the year when the sun shines brightly. The sun's light is very severe on paint materials and, for that reason, synthetic paints are proving popular in Australia. The use of synthetics in a paint may increase its life 100% in a climate like this. The paint trade and the public in this country appear to realise this for our sales of synthetics in Australia in 1936 was four times as great as in 1935. ("Dr. Krumbhaar's Visit" 1937)

Sunday Reed must have had some success in finding paint for Nolan, for on October 31, 1942, she lists in her diary a work titled "Lightning" as having been received. This entry is a good match with a work called *Lightning, Dimboola,* painted on a sheet of paper torn from a spiral-bound sketchbook with an inscribed date "31-10-42" (fig. 18). Whether this work is painted with an alkyd is not known, not having been analytically tested, but three closely related works (not noted in Sunday's diary), all of them, like *Lightning,* torn from a similarly sized spiral-bound sketchbook and dated within days of each other, have been tested. The paints on the three works—*On the Beach* (October 28, 1942; fig. 19), *Nostalgia for the Sky* (November 1, 1942; ; fig. 20), and *Farmer, Dimboola* (November 2, 1942 fig. 21)—have all been identified as alkyd paints (see appendix I).

Applied to thick but uncoated paper, some of the alkyd colors on the paintings, particularly the powerfully tinting pigment Prussian blue, have bled through to the back of the sheet, but otherwise the paper appears to be little affected by the solvent paints. Although sharp cracks are evident in the thickest layers, the stability and flexibility of alkyd paints are well demonstrated by the excellent condition of the works. Paint layers must have dried quickly but perhaps not before some transfer of color onto the next sheet occurred. The works from November 1 and 2, though different in subject, share remarkable compositional similarities, as if they are based on transferred images. The metamorphosis of the alienlike creature in *Nostalgia for the Sky* into the

FIGURE 18 (top left)
Sidney Nolan, *Lightning, Dimboola*, October 31, 1942. Lacquer on paper, 23.5 × 26.5 cm (9¼ × 10½ in.). Heide Museum of Modern Art. Bequest of John and Sunday Reed 1982.

FIGURE 19 (top right)
Sidney Nolan, *On the Beach*, October 28, 1942. Alkyd and oil on ivory wove paper (page from spiral bound sketchbook), 24.7 × 26.9 cm (9¾ × 10⅝ in.). Art Gallery of NSW. Purchased 2013.

FIGURE 20 (bottom left)
Sidney Nolan, *Nostalgia for the Sky*, November 1, 1942. Alkyd and oil on ivory wove paper (page from spiral bound sketchbook), 24.6 × 27.1 cm (9¾ × 10¾ in.). Art Gallery of NSW. Purchased 2013.

FIGURE 21 (bottom right)
Sidney Nolan, *Farmer, Dimboola*, November 2, 1942. Alkyd on ivory wove paper (page from spiral bound sketchbook), 24.7 × 27.3 cm (9¾ × 10¾ in.). Art Gallery of NSW. Purchased 2013.

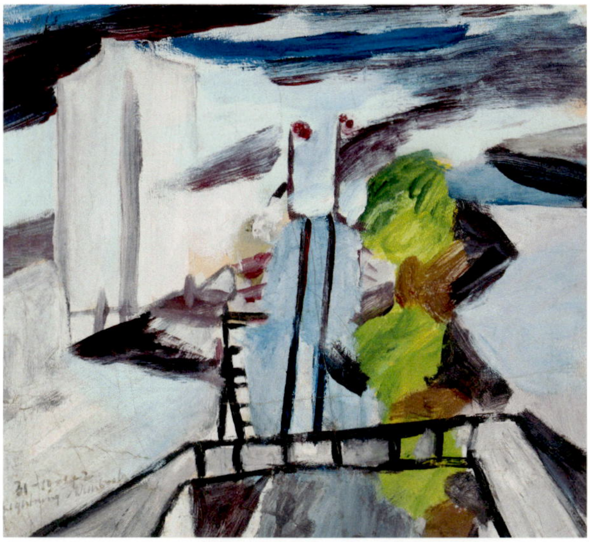

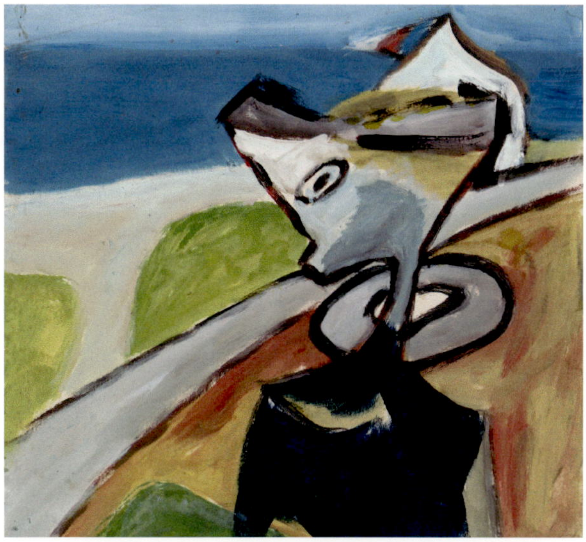

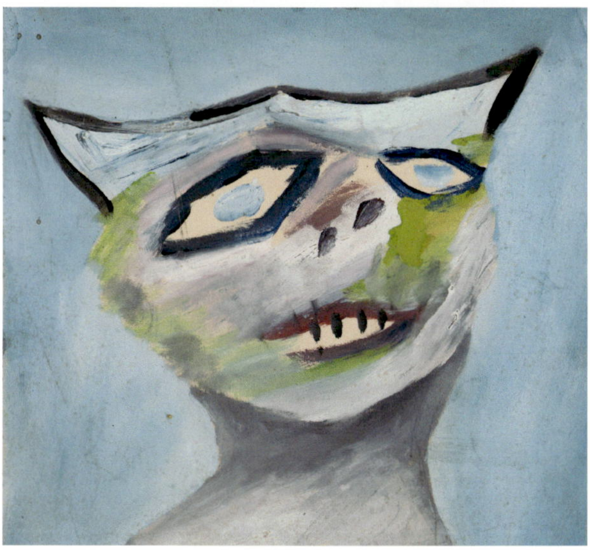

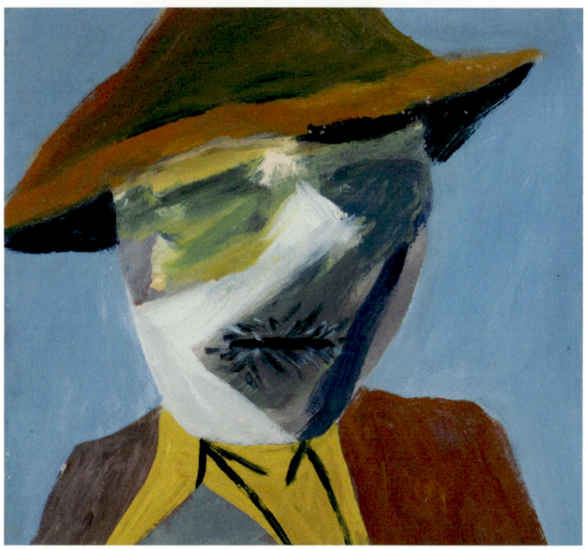

FIGURE 22
Sidney Nolan, *Dimboola*, December 5, 1942. Alkyd on cardboard, 62.0 × 74.7 cm (12⅜ × 29⅜ in.). Heide Museum of Modern Art. Bequest of John and Sunday Reed 1982.

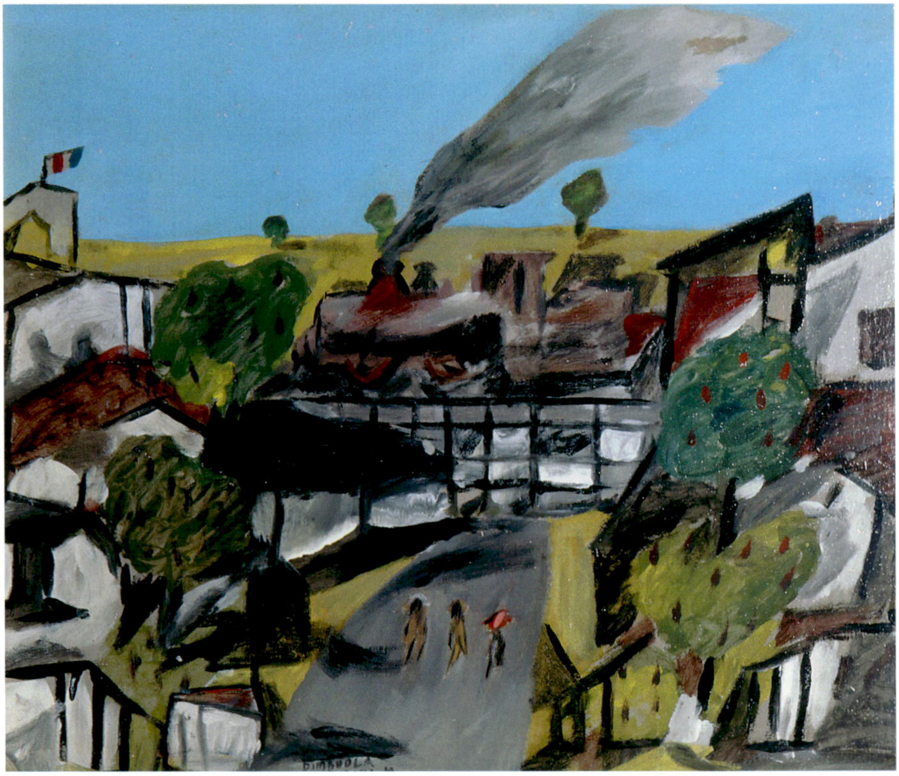

wizened-lipped figure of *Farmer, Dimboola* the following day is one of the strangest of Nolan's compositional progressions.

Other 1942 paintings by Nolan that have been analytically identified as alkyds include *Water Wheel, Luna Park* (1942) and *Dimboola* (1942; fig. 22). It would appear, therefore, that Nolan and Sunday were successful in their search for paint, and perhaps they found more than just the three primary colors along with black and white that Nolan initially requested. Many of these works, for example, feature green paint. Whether this was a purchased color or mixed by Nolan is not easily ascertained; green household paints were generally manufactured by mixing Prussian blue with chrome yellow rather than using a pure green pigment.

In any case, the task of obtaining paints at this stage of the war must not have been easy. Dulux and other brands of alkyd paint were specified as essential for war armaments and vehicles, and from 1942 paint makers were prohibited from manufacturing alkyd paints for civilian use (Australian Government 1942). Although there were no restrictions on the sale of preexisting supplies, by 1942 synthetic paints of all types were becoming very hard to find.

An additional pressure on the supply of alkyds during the war was that dimethyl phthalate, a key starting compound of the alkyd resin, was also an effective mosquito repellent. Large amounts were therefore diverted from paint manufacture into insecticide production, to help protect troops from mosquito-borne malaria (Boland 1944–45). Dimethyl phthalate (code named "Mary") was manufactured for the Australian Army by a Sydney paint company, Slade, from late 1943 (Todd 2007, 26). Filling small, colorless bottles with three finger dimples, it was issued to all military personnel in tropical Australia, New Guinea, and the Pacific. Remarkably, just such a bottle was found among Nolan's paints in his studio in Wahroonga, Sydney, and the still-liquid contents were identified as dimethyl phthalate. Nolan's bottle was manufactured for the

FIGURE 23
Sherwin-Williams Solves Wartime Problem, Advertisement in *The Decorator & Painter for Australia & New Zealand*, January 17, 1942, 15. Mitchell Library, State Library of New South Wales.

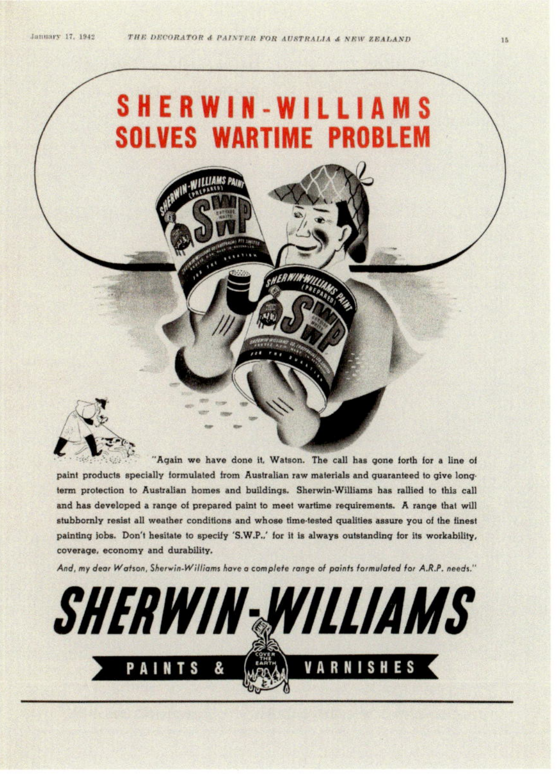

Australian Department of Defence, as evidenced by the D/broad arrow/D embossed on the base, and probably issued during his army service. Being presented with a bottle of "Mary" would likely have been a strong indication to any Australian soldier in the period 1943–45 that departure for the battlefields in New Guinea or the Pacific was imminent. Perhaps it was so for Nolan also.

Household paint manufactured for domestic use in Australia after 1942 reverted to traditional ingredients that were more plentiful and locally available (fig. 23). These were principally linseed oil mixed with a white pigment made from 90 percent lead white and 10 percent zinc white, making a paint called master painter's white. On February 4, 1943, Nolan wrote to Sunday from Dimboola about white and gray paints for priming boards:

> I got some white dulux primer which is different from the gray but it is a necessity to have plenty of it. I brought a tin of master painters white, liked the name probably, 90% lead & 10% [*sic*]. Old stock, there is no zinc now being put up. You can try it, if satisfactory I can get another ½ gal. It's not the same as titanium & zinc & fresh eggs[4] I know, but we will have to allow some margin to get paint at all.

Paintings from Nolan's army period often utilize a brilliant white matte priming layer that, when left in areas unpainted as part of the final image, provides a contrast with the high-gloss finish of the colored areas. This important variation in surface gloss is visible only on those paintings that remain unvarnished, such as *Bathers* (1943; fig. 24). The priming layer on this painting—visible as the white stripes in towels laid on the decking and shaken in the air by a bather, and in the diving tower and tents on the beach—is a mixture of pigments, including lead white (basic lead carbonate), chalk

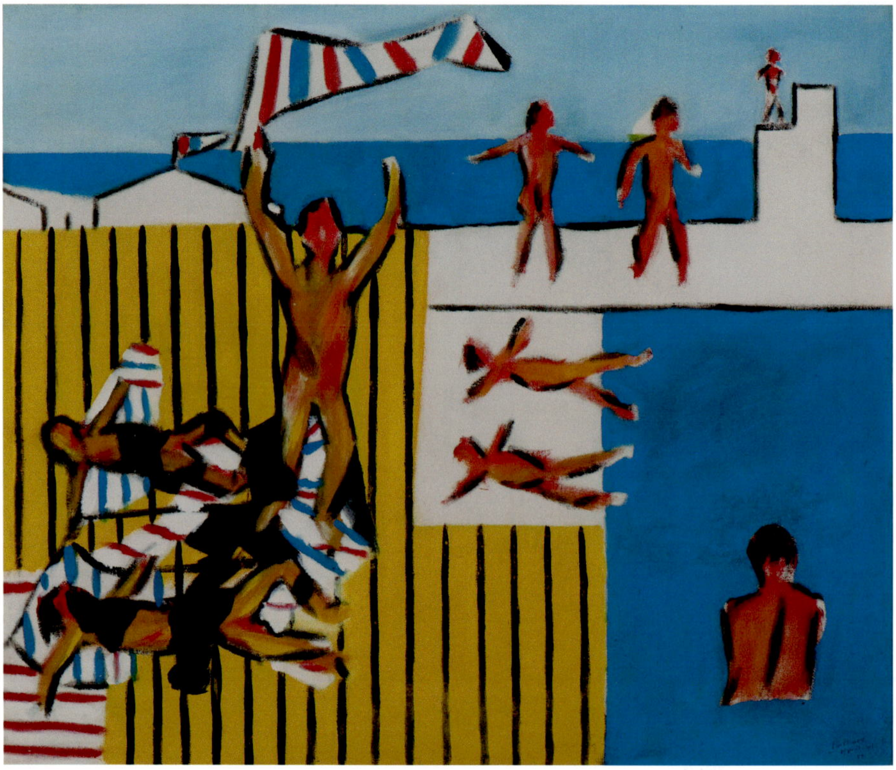

FIGURE 24
Sidney Nolan, *Bathers*, April 12, 1943. Ripolin oil enamel on cotton canvas, 62.9 × 75.5 cm (24¾ × 29¾ in.). Heide Museum of Modern Art. Bequest of John and Sunday Reed 1982.

(calcium sulphate), barites (barium sulphate), kaolin (aluminum silicate), and zinc white. Other paintings, such as *Dimboola* (see fig. 22), have a dark background achieved by priming with a gray paint, in this instance an alkyd mixed with kaolin, titanium white (titanium dioxide), barites, and zinc white. Nolan's priming layers applied to cardboard and Masonite supports in this period are a complex mix of white and filler pigments, suggesting the use of different types and brands of household paints.

Wartime Supports

At this stage of the war, Nolan and Sunday Reed faced considerable difficulties in obtaining not just paint, but also canvas and boards to paint on as well as brushes, typically pig's bristles imported from Russia ("Prohibition and Restrictions on Paints and Brushes" 1942). In October 1942 Nolan wrote to Sunday about supports: "To be practical. I have hunted around here but there is no wood to paint on, the lads have burnt any there might have been, so it is a good plan to send up whatever you can find in the way of cardboard for a start." A 1943 work by Nolan, *Untitled (St. Kilda),* was painted on either the cardboard insert from a packet of Ilford panchromatic film plates or a piece of cardboard cut from the box itself. The back of the painting retains the label from the film and instructions for developing. Some successful scrounging of materials directly from the army must also have occurred, judging from a faint ink stamp reading "Dept. War Organization of Industry" visible on the wooden strainer[5] for *Self Portrait* (1943; fig. 25). Because the cloth was too narrow to wrap around the sides of the strainer, the canvas, of rough hessian, was attached to the front using a combination of drawing pins and nails, which were painted over by the artist during the making of the work. This

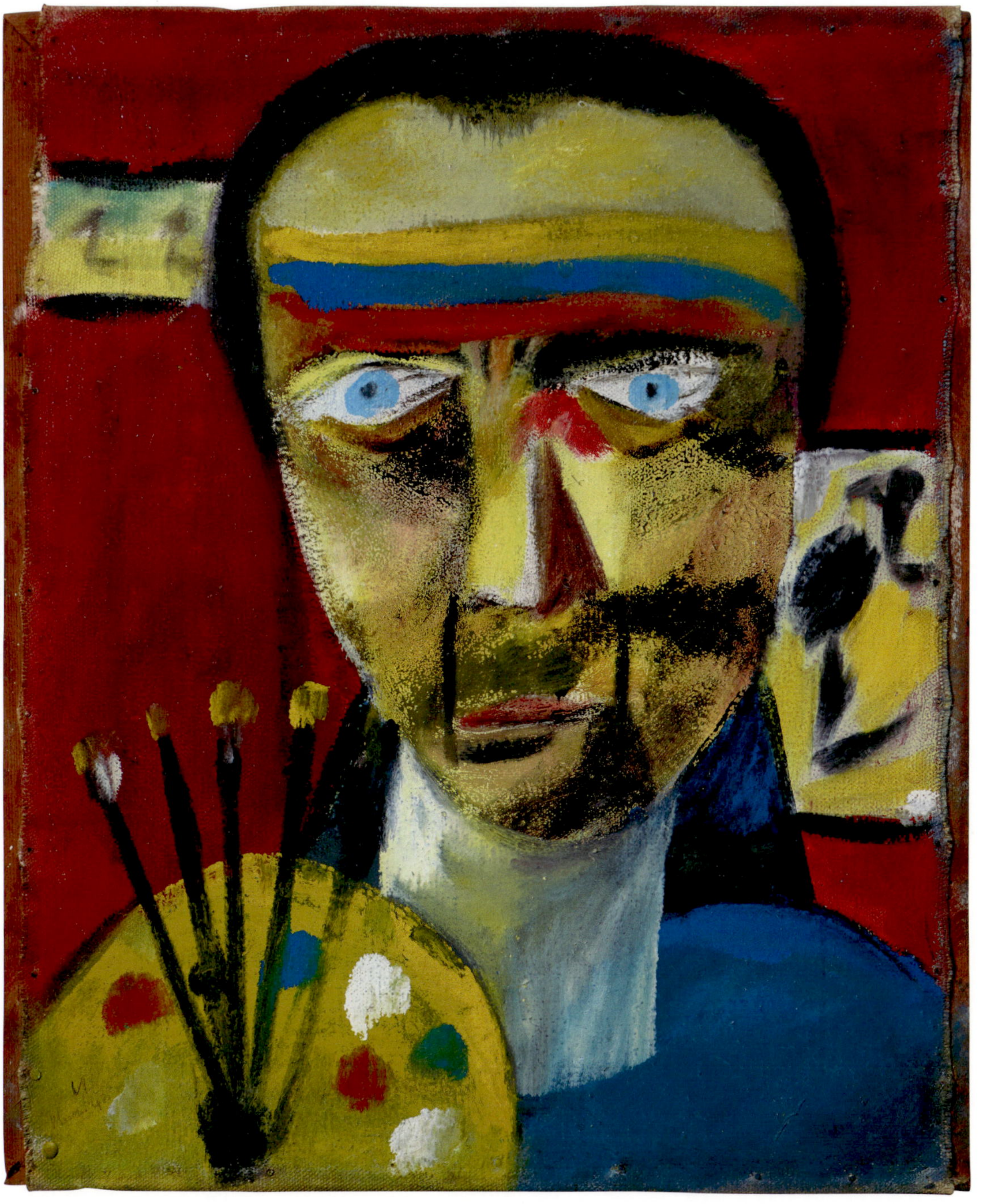

FIGURE 25
Sidney Nolan, *Self Portrait*, March 1943. Ripolin oil enamel on hessian canvas, 61.0 × 52.0 cm (24 × 20½ in.). Art Gallery of NSW. Purchased with funds provided by the Art Gallery Society of NSW 1997.

unusual means of attaching the canvas has been retained and is visible with the work in the frame.

The Department for the War Organization of Industry (WOI), established in October 1941, assigned war production to industries while maintaining a sustainable level of civilian production (Australia, Dept. of War Organization of Industry 1943). Part of its strategy was to reduce varieties of products considered a drain on resources during wartime. The department issued a decree simplifying clothing styles, for example—prohibiting trouser cuffs and shirt pockets—to reduce material use, since linen was impossible to procure during the war and cotton was second only to food in being rationed. The restrictions on civilian use of cloth had a significant impact on artists during the period and led to the common substitution of cardboard and builder's boards for traditional canvases. Muslin, too, which Nolan had become partial to in 1942 as a textured surface when adhered to the smooth side of Masonite or cardboard (a technique known as marouflage), was hard to obtain owing to its assigned use as dressing for wounds (Kubik 2006).[6] Nolan, as always resourceful, was able to find some precious stock and post it back to Sunday Reed so she could prepare his supports for painting.

> I posted Mac's novel today with six yards of muslin wrapped around it. Doctors certificates are required here for muslin but the man at the canvas place is anti Redman[7] so he gave me the six yards. 7½^{d} [pence per] yard. It doesn't look very suitable but as its all I can get I took it in any case. Have to make cheese with it if it won't marouflage. (Nolan to Sunday Reed, c. Autumn 1943)

Builder's wallboard made from pulped wood (known as hardboard) such as Masonite had become Nolan's preferred painting support in late 1942, but it grew increasingly difficult to obtain as it was diverted to the battlefront for the building of shelters. Masonite was a new material in Australia in the early 1940s. Manufactured from 1938 at a factory in Raymond Terrace, north of Sydney, under license from the American Masonite Company, Australian Masonite was made from indigenous hardwoods, particularly trees from the Casuarinaceae family (commonly called she-oak), growing in the swampy land around the factory. This timber source produced a distinctively dark-colored hardboard, easily distinguished from the pale boards manufactured in Europe and the United States from softwoods. Unlike modern composition boards made from wood fibers, such as chipboard and medium-density fiberboard, Masonite was made without the addition of adhesives. Instead, the natural lignin in the wood adhered the particles together under huge amounts of steam and pressure.

The advantage of Masonite to artists lay in its large size and portability. Sunday Reed's success in finding two big sheets in October 1942 elicited a celebratory letter from Nolan in which he outlined specific instructions for how best to cut it into many useful smaller pieces without wastage.

> Getting the Masonite was a stroke of good management, I thought it was extinct, went out with the Kiwi. Now that you have it, it is going to be a job cutting it, or sawing probably. About cutting, if it is 12 ft × 4 ft which I hope you said, measure 18" eight times along the 12 ft side which will give you 8 pieces 18" × 4 ft when cut. Cut these in half along the 4 ft side which will give 16 pieces 18" × 2 ft, or about the size of the Herald which is not a bad size & shape.[8] If a smaller size were wanted at any time that size would again cut in half along the 2 ft side making 18" × 12" like the Herald folded in two. That's a fair bit for a start. I think we'd better leave the other sheet uncut for the moment, in case of a magnum opus, but about the size of a newspaper is magnum enough at present. (Nolan to Sunday Reed, October 27, 1942)

Masonite is also distinctive for having two different sides, one smooth and the other with a pressed wire pattern giving the appearance of a woven canvas when primed and painted. Early on, Nolan typically used the rough side, for example in *Wimmera Landscape* (see fig. 15), or he glued muslin cloth to cardboard to imitate canvas texture, for, Nolan said, "the Ripolin washed in the muslin like watercolour" (Nolan to Sunday

Reed, January 12, 1944). By mid-1943, however, he consistently painted on the smooth side of Masonite or cardboard without muslin, embracing the board's inherent flatness. The combined use of these boards with the glossy household paints that flowed out in an even layer produced paintings with the smooth texture of porcelain. The unique surfaces of his paintings earned Nolan the nickname of "Sid Linoleum" among the Heide circle (Barbara Blackman, conversation with author, 2016).

Nolan's Ripolin

Restricted paint supplies spurred Nolan and Sunday Reed to scour hardware and paint shops in their determination to find the materials for painting. The most miraculous success was a supply of high-quality Ripolin enamel paint. In a letter of October 27, 1942, Nolan had suggested to Sunday that they begin to look for some: "There must be a stock of Ripolin somewhere in the country, there was a Ripolin House in Sydney, but Bert [Albert Tucker] said they were no go. If you could get their address from him I will give them a fling anyway. However Dulux it is right now." Success in their quest was announced within a couple of months, with Nolan writing: "You must have had a long job with priming all the masonite & now the ripolin has arrived" (Nolan to Sunday Reed, December 1942, 6). On January 31, 1943, Nolan, now living in Nhill, wrote Reed to say that he had the Ripolin in hand: "I collected the canvases this morning from the station this morning [*sic*] but I am almost frightened to open the ripolin in case it disappears in a cloud of fumes."

In 1942 it was remarkable that Ripolin was available anywhere in Australia, and it is most likely that Nolan's first consignment was prewar stock. The probable source was, as Albert Tucker had suggested, Ripolin House in Sydney. The Australian distributor for Ripolin paint from at least the 1910s was Lindsay McCormick, who owned Ripolin House at 25–27 Clarence Street, Sydney, from 1930 (Sands 1930). He was still in business at the same address in 1939 offering Ripolin paint at a discount, which suggests that good stock remained in Australia at the outbreak of war (*DPA&NZ*, March 10, 1939).

Although Nolan probably already knew of Ripolin through his own commercial art background, he acknowledged that his search for a supply of the paint was a result of Picasso's well-documented endorsement, as described by Gertrude Stein in *The Autobiography of Alice B. Toklas* (1933, 173):

> It was in that year [1914] that he [Picasso] began to use ripolin paints instead of the usual colours used by painters. Just the other day he was talking a long time about the ripolin paints. They are, he said gravely, la santé des couleurs, that is they are the basis of good health for paints. In those days he painted pictures and everything with ripolin paints as he still does, and as so many of his followers young and old do.

As Nolan explained in a 1962 interview, "I read that Picasso had said that Ripolin was a healthy paint, and I thought, well, why should I bugger around and torment myself and a wise man says it's healthy paint. Okay, so Ripolin it had to be" (B. Smith 1962). It is likely that Picasso's enigmatic "la santé" claim concerned the brilliance and permanence of Ripolin's colors, but the paint was also a healthy choice for artists. Instead of the toxic lead white (basic lead carbonate) pigment used in most Australian-made household and commercial paints of the period, Ripolin was based on zinc white (zinc oxide). The Australian Ripolin distributor, McCormick, went still further in his healthy claim for the paint, writing in 1914 to the Nepean Hospital to suggest its use in the painting of the new hospital building, quoting a test made by Dr. F. Baseman of the University of Leiden that demonstrated the paint's germicidal properties ("Our Hospital" 1914).

Nolan quickly went to work with his new paint medium, writing excitedly to Sunday:

> You should have been here for the unveiling of the ripolin. Perhaps it is going to take longer to paint good pictures now, it seems like it when you have to dig deeper & perhaps it will be a while before I really know the paint. But the paint itself will take all the discipline in the world. Some day I feel I know what paintings will come from it. (Nolan to Sunday Reed, February 2, 1943)

This shift from Dulux to Ripolin in February 1943 is also documented in Sunday Reed's diaries, where she notes the media used in Nolan's paintings. Invariably from December 9, 1942, through to January 1943 Dulux is specified, but from February 1943 onward Ripolin is the only paint medium recorded, up to the last diary entry, of April 17, 1943.[9] Nonetheless, art historians and writers on Nolan's work have generally held that he was using Ripolin paint prior to 1943. Nolan's biographer Brian Adams stated that Nolan used Ripolin in 1940 when painting on the slate roofing tiles (Adams 1987, 50). Elwyn Lynn catalogued works by Nolan painted with Ripolin as early as 1941, including *Bird* (Lynn 1979, 24), which has subsequently been analytically identified as nitrocellulose (see appendix I), and Lou Klepac suggests Nolan was using Ripolin prior to entering the army in 1942 (Klepac 2007, 79). Janine Burke, in discussing Sunday Reed's diary entry of February 5, 1943—"painted first Ripolin canvas"—takes it to signify that Sunday is marking the occasion of Nolan's first Ripolin painting *on canvas*, not that it was his first painting *using* Ripolin (Burke 2005, 203).

The mistaken practice of assigning the medium description Ripolin to all of Nolan's works in gloss paint from the 1940s overlooks his wide-ranging prior use of locally made alkyd and nitrocellulose paints. The misuse of the Ripolin description is, however, in part a problem that Nolan created himself. When interviewed in 1962, he was asked, "When did you start using Ripolin, then, Sid. Remember?" Nolan replied, "Oh yes, before I went into the army, just before, probably '42. More or less the first Ripolin I had went into the army with me, I think, about a year before, I suppose." Asked where he obtained his stock, he replied, "From London" (B. Smith 1962).

Just as Nolan's letters and Reed's diary entries belie the notion that he was using Ripolin prior to 1943, it is also highly unlikely that Nolan was able to import paint directly from London during the war. Why he made these assertions in 1962 is not clear. At the time, Nolan was seeing his own paintings from the early to mid-1940s for the first time since he had left Melbourne in 1947. It may therefore have simply been a poor or inexact recollection of dates and events, or even a deliberate attempt to obscure Sunday Reed's close technical assistance in the wake of their unhappy breakup in 1948. Perhaps Nolan merely sought to deflect scrutiny of his processes and materials, wishing to keep an elusive chemical magic hidden by "leading the hunter away from its young and nest," as Saul Steinberg described Nolan's strategy (quoted in C. Nolan 1967, 100).

A transitional painting is *Boy in Township* (1943; fig. 26), a striking instance of Nolan using one of his own paintings as background for another work. Originally a Wimmera landscape with a pink sky, it was rotated a full 180 degrees and repainted with a figure and blue sky.[10] The upside-down buildings allow the lower painting to be

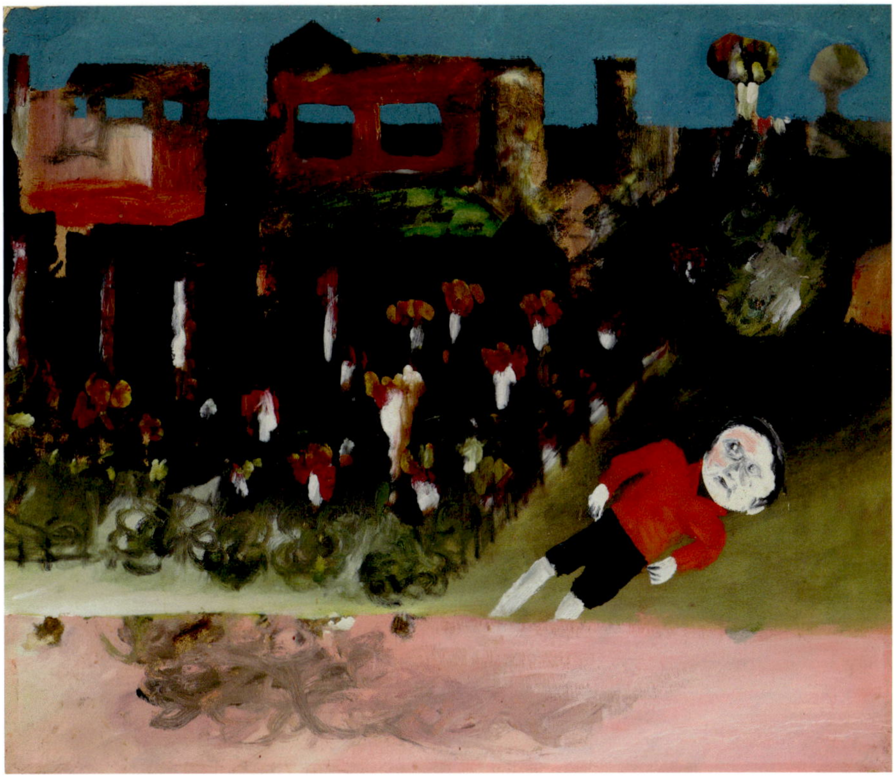

FIGURE 26
Sidney Nolan, *Boy in Township*, 1943. Ripolin oil enamel on cardboard, 61.0 × 72.5 cm (24 × 28½ in.). Art Gallery of NSW. Purchased 1962.

seen through the walls and roofs, and the pink sky of the first painting is retrofitted to a foreground river in which the boys feet disappear. The landscape has become a strangely rotated world not unlike one viewed while lying flat on the ground. While the analysis of the upper paint layer is consistent with the use of Ripolin, the pigment in the pink paint of the lower painting (originally the sky) was identified as carmine, a natural red dye obtained from cochineal, which was mixed with titanium white. Carmine is not found in any of the cans of Ripolin paint left in Nolan's Wahroonga studio (see following chapter for detailed description of pigments identified in Nolan's cans of Ripolin paint). It is probable that the lower painting predated the arrival of the Ripolin and he then used Ripolin for the rotated repaint after it arrived in January 1943. In late 1942 Nolan discussed adding dyes to white paint because of the difficulty of combining premixed tints:

> The other shades, funny word shade isn't it, red, pinks, green are difficult to mix with each other, & if you have any choice are best to avoid. But as before white is important. Either Dulux or Dynamel & the Duco firm also put out a "Duco white lacquer" & white lacquer undercoat which all serve as a base for strong color. If you are going to get a bit of a hoard together darling it would be good to see Deans about more aniline dye. They had some before which came from England. Ask if they have the three colors, red, blue & yellow again soluble in spirit. And also three soluble in water, although the spirit ones are really all that is necessary. And black. But as strong as they have got, they probably call them chrome yellow, crimson or scarlet & prussian blue. (Nolan to Sunday Reed, October 27, 1942)

Although Prussian blue (iron ferrocyanide) and chrome yellow (lead chromate) are not dyes but pigments, they were often used as paint tinters for their fine particle size and strength of color. Red-colored tinters were usually dyes, either natural, such as carmine, or synthetic coal-tar reds (see sidebar on p. 18: Coal-Tar Colors).

Inscribed "March 1943," *Self Portrait* (see fig. 25) dates to soon after the arrival of the Ripolin from Sunday Reed in late January. This painting is probably the work listed in Reed's diary as "Head Nhill," which she received on March 22 and describes as "Ripolin on hessian." Despite being in the new paint, it retains the three primary colors along with black and white that Nolan had favored with the Dulux. However, the pigments in this painting—zinc white (zinc oxide) and barites (barium sulphate) mixed with the colored pigments Prussian blue, toluidine red (PR 3), chrome yellow, and carbon black—are all good matches with the pigments and the oil medium identified in cans of Ripolin from Nolan's Wahroonga studio.

However, a flat artist's palette used for mixing artist's oil colors, as depicted in *Self Portrait,* would have been impossible with Ripolin: the paint would simply have poured off the edges. Ripolin's tendency to flow uncontrollably is evidenced at the hairline of the self-portrait, where the black paint has dripped into the forehead, in this instance softening the line between hair and skin. Another unusual feature of the Ripolin paint on this work is the resistance, separation, and pooling of the black paint where it was applied over the lighter paint of the face to suggest shadows. This problem is caused by the surface drying of enamel paints (related to the use of cobalt driers in the paint) and their smoothness; once dry, the paint resists liquids, even another Ripolin paint. Serendipitously, the broken-apart field of black paint softens the shadow; what might be considered paint defects have been turned into convincing assets with typical Nolan bravado, the paint itself asserting its own rules of application.

Thus Ripolin paint proved both inspiring and difficult for Nolan, in that it required new ways of working. Because the paint was slower to dry than the nitrocellulose and alkyd paints he had used previously, he had to leave paintings drying flat for days in front of a small heater—a frustrating necessity, as he preferred to work more quickly. Of his first painting in Ripolin he said, "Finished one & there appears to be the beginning of a glow that Ripolin can give. But I bit my teeth on lacquer & the changeover is really quite considerable" (Nolan to Sunday Reed, February 9, 1943).

The Ripolin paintings he made while in the army, limited as they were by circumstance, set Nolan on a technical pathway that he continued well after the war finished and other media, including artist's paints, were once again procurable. During the wartime shortages of materials Nolan had written that he would like some artist's tube paints: "still a hankering after oil-color but I really feel it belongs with Rembrandt & Cezanne—or after the war" (Nolan to Sunday Reed, c. October 1942, 3). It seems, however, that the advantages he found in the cans of Ripolin inspired a continuing commitment to this paint for at least another decade. (For more on the history and physical make-up of Ripolin, see chapter 3.)

Sheffield Exhibitions

In addition to procuring materials, priming supports, and keeping records during the period of Nolan's army service, Sunday Reed arranged several exhibitions of his work in the window of Sheffield's, a general store in Heidelberg. The first Sheffield exhibition opened in July 1942, and a second one, in July 1944, was photographed (fig. 27), showing Sunday Reed leaning on a car near the display). As Nolan observed to Sunday about the

FIGURE 27
Photographer unknown, Second Nolan exhibition at Sheffield's, Heidelberg, Victoria, July 1944. Heide Museum of Modern Art.

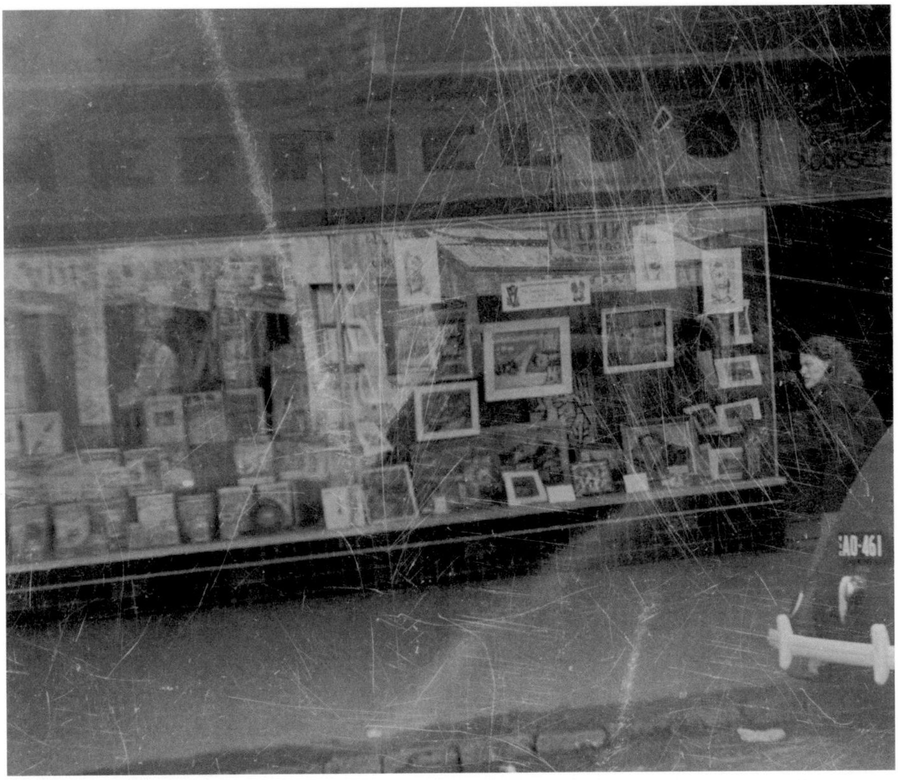

1942 exhibition, perhaps recalling his work making displays for Fayrefield Hats: "It is practical to show pictures in windows, our equivalent of a market place" (July 19, 1942).

Most of the works in the 1944 display were shown with simple light-colored painted frames, which Sunday herself may have crafted. Letters to her from Nolan are sometimes addressed "Dear framemaker," and he commented, "Frames are going to be a big point. The idea you had of putting the canvas flush with the frame sounds good" (Nolan to Sunday Reed, March 29, 1943). This framing trend, which opposed the traditional approach of providing a recessed window for the representation of three-dimensional space, helped to emphasize the flatness of Nolan's picture plane.

Nolan and Sunday were also experimenting with a modeling compound called Nucraft. Sunday designed a frame with "Nucraft, string, black pepper & old paint" for *Apples* (c. 1941), no longer present on the painting, and Nolan modeled a gun from "Nucraft & sack & twig & painted lacquer & dye." The gun was exhibited in the 1942 Sheffield exhibition, but it fell apart and had to be removed.[11] Unfortunately, the composition of this modeling material is unknown, and the frame and gun are no longer extant for analysis. A sculpted gun in 1942, however, is an intriguing clue to Nolan's interests, presaging as it does a group of paintings that he would develop after deserting from the army in 1944, inspired by the story of the nineteenth-century bushranger Ned Kelly.

3 Postwar Kelly: 1945–1947

In July 1944, fearing imminent departure for the battlefront (perhaps warned by the issuing of the bottle of mosquito repellent), Nolan deserted from the Australian Army. He went into hiding, living both at the Reeds' house, Heide, and in a bedsit in Parkville, Melbourne. He continued to paint throughout this time. Photographs taken by Albert Tucker at Parkville show Nolan's technique of working almost flat (fig. 28), a strategy that prevented the Ripolin from running uncontrollably down the surface of the painting. Sunday Reed must have questioned this practice, for on May 18, 1943, he had written reassuring her that "all good painters work on the floor, Sun."[1]

During his illegal AWOL status, Nolan became interested in the story of the notorious bushranger Ned Kelly. The nineteenth-century outlaw was known and admired by many for his rebellious stand against the police, who hunted him over several years. To protect himself from the bullets of his pursuers, Kelly crafted body armor from the iron of a plow moldboard, consisting of a crude cylindrical helmet and upper body plates held together with leather straps and bolts. Nolan's representation of this helmet as a black square with a horizontal slot opening over the eyes, and of the body armor with bolts at the shoulder, became the painter's powerful visual shorthand for the historic and folkloric figure.

Nolan researched Kelly's personal insights into the tragic circumstances of his life, including poverty, police harassment, and his own criminal behavior, leading ultimately to his conviction for murder and subsequent death by hanging.

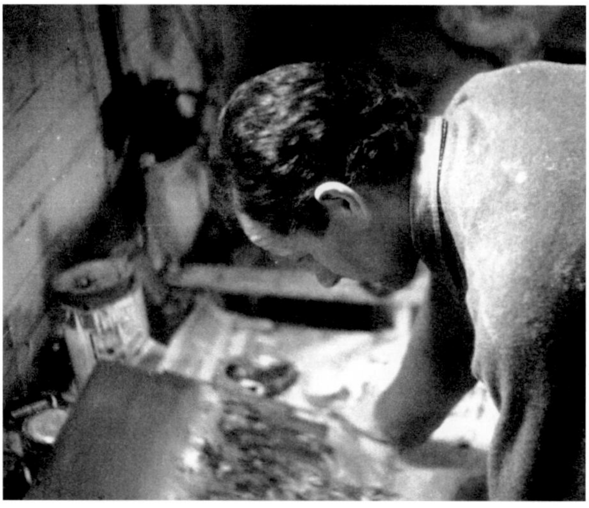

FIGURE 28
Albert Tucker, Nolan looking at *Dimboola* in Parkville studio, September 1944. Papers of Sidney Nolan, National Library of Australia, MS 10245, Series 2, File n. 151.

FIGURE 29
Albert Tucker, Nolan's worktable in the dining room at Heide, c. 1945. Silver gelatin photograph, 30.4 × 40.5 cm (12 × 16 in.). Heide Museum of Modern Art. Gift of Barbara Tucker 2001.

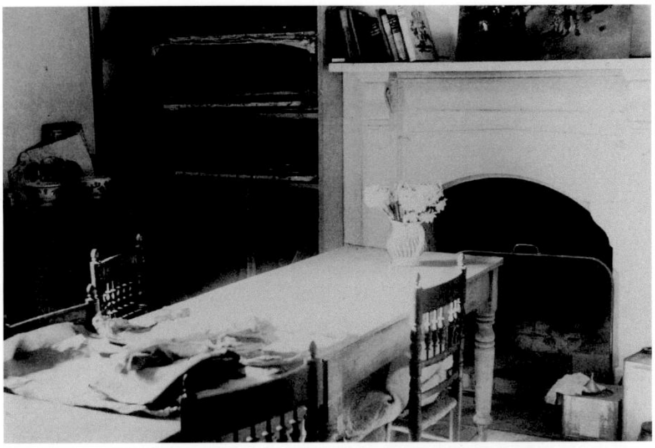

While Nolan was working on a suite of Kelly paintings on cardboard and Masonite panels in the dining room of Heide, a visiting friend, Barrett Reid, commented that "the dining room was a studio, with tins of Ripolin, bottles of oil, a scrubbed long table" (quoted in Burke 2005, 256). A photograph from 1945 shows the table cleared of paintings, but the shelving beside the fireplace appears to have stacked paintings interleaved with newspaper, and one work is propped up on the mantelpiece (fig. 29). Several of the Kelly paintings retain the impression in the thicker areas of white paint of having been in contact with the back of another sheet of Masonite while the paint was still soft. On occasion, the newspaper used to separate the panels has also transferred to the paint.

Barrett Reid's comment about bottles of oil in addition to the cans of Ripolin is of interest as it has not been previously noted that Nolan was working with anything other than Ripolin paint on the Kelly paintings. The addition of oil is supported, however, by a photograph taken by Albert Tucker outside Heide in about 1946 showing a number of large glass jars filled with liquid, which Burke (2005, 256) identified as linseed oil left in the sun to bleach.

Doerner's *Materials of the Artist and Their Use in Paintings* (1934, 105) extolled the virtues of sun-thickened linseed oil for painting: "It gives colours an enamel-like character and permits, despite its viscidity, a great amount of technical freedom. Since it has already absorbed oxygen, it dries more quickly than ordinary linseed oil. . . . It does not crack and is very elastic." To obtain this thickening, Doerner suggested pouring the oil into a shallow dish, which allowed for maximum exposure to the air (oxidation) as well as to light and heat (polymerization). The oil photographed by Tucker, however, is in narrow-necked bottles, not open saucers, which could hardly have offered enough surface area to allow the oil to oxidize.

Another technical guide gives a recipe that may be a better fit with Nolan's process. *The Artist's Handbook of Materials and Techniques* by Ralph Mayer, published in 1940, describes a method for making sun-refined or sun-bleached oil. Mayer notes that traditional techniques, such as that suggested by Doerner, using sun and air to oxidize, polymerize, and bleach the oil caused excessive thickening and yellowing. He recommended instead using a narrow-mouthed jar, which, while allowing exposure to light, thus promoting bleaching and polymerization, would reduce yellowing of the oil by inhibiting oxidation. Pre-polymerized linseed oil—heated in the absence of oxygen—was used broadly in the paint industry and was called *stand oil*. Mayer also suggested adding water to the oil in equal proportions and shaking the two together daily, leaving them exposed to the sun for a few weeks. This would allow the water-soluble mucilage and impurities to separate from the oil. Oil thus prepared was suitable for use as a clear varnish or glaze and for painting-medium purposes. One can imagine that stand oil,

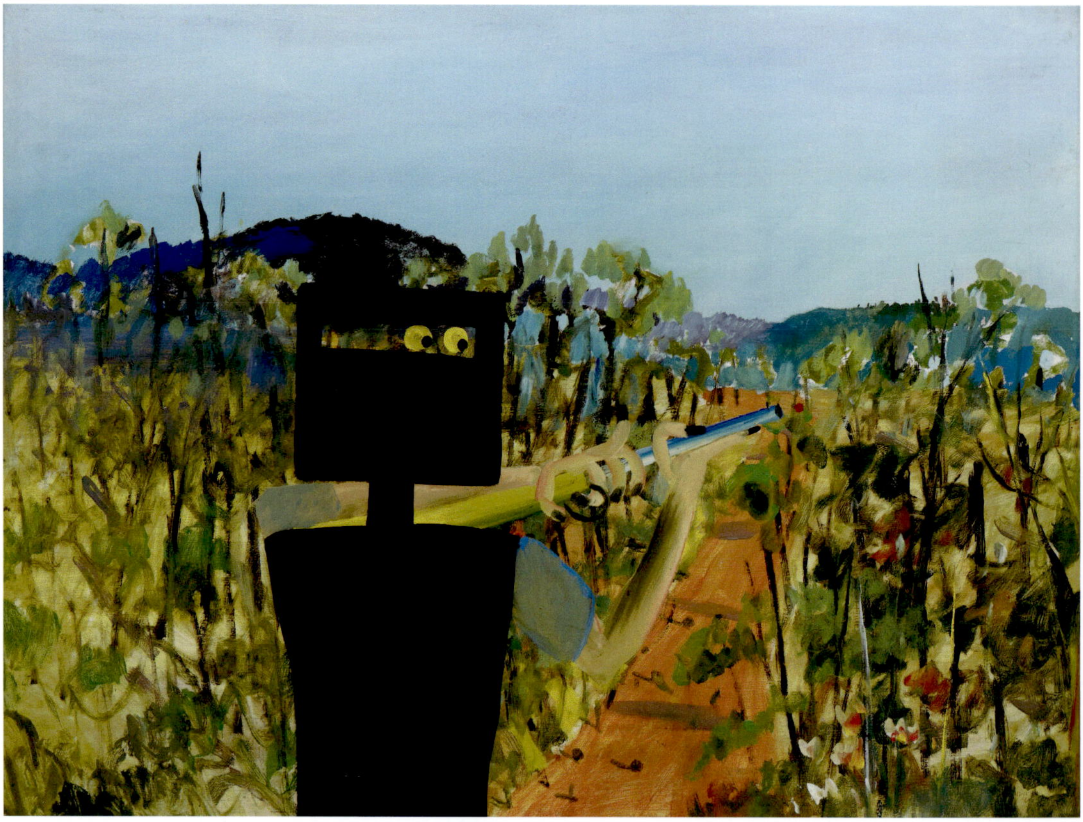

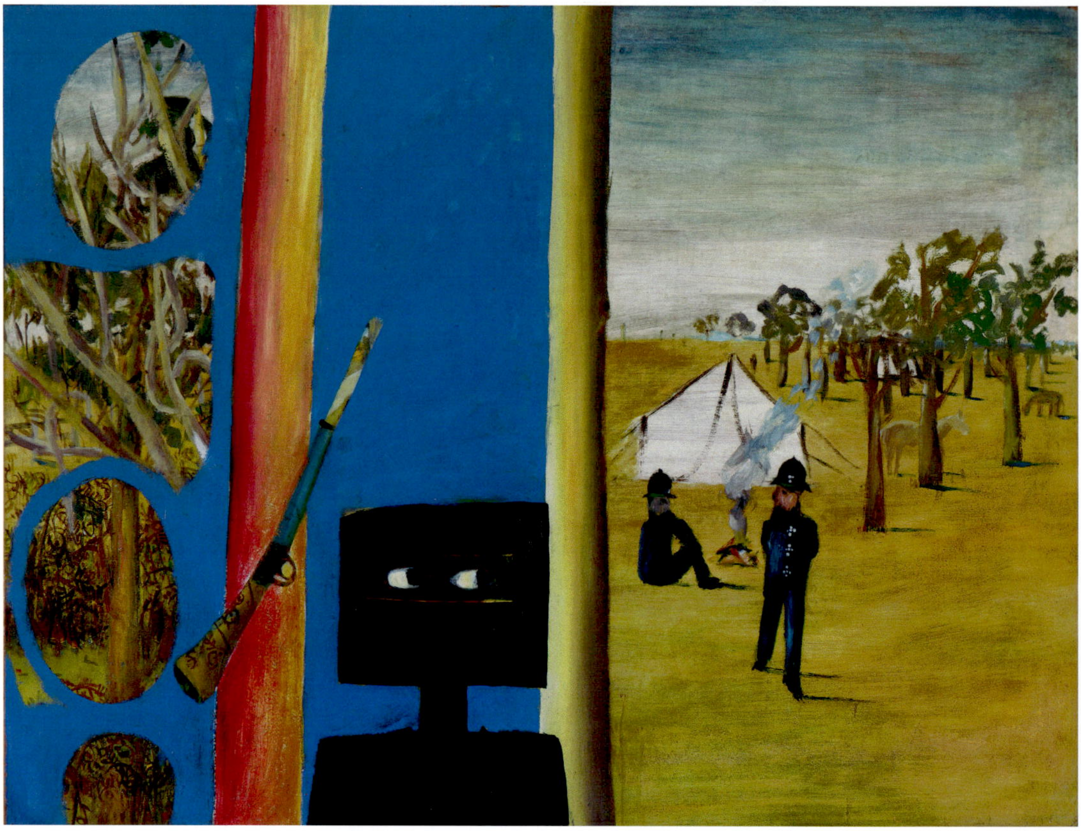

FIGURE 30
Sidney Nolan, *First-Class Marksman: "While in the ranges they indulged in rifle and revolver practice to such an extent that all of the four mates were first-class marksmen with any kind of fireman* [*sic*]*,"* December 1946. Ripolin oil enamel and alkyd on Masonite hardboard, 90.2 × 121.2 cm (35½ × 47¾ in.). Art Gallery of NSW. Purchased with funds provided by the Gleeson O'Keefe Foundation 2010.

FIGURE 31
Sidney Nolan, *The Camp*, 1946. Ripolin oil enamel and alkyd on Masonite hardboard, 89.7 × 121.5 cm (35¼ × 47⅞ in.). Art Gallery of NSW.

fast-drying, self-leveling, and with good longevity, would have been attractive to Nolan. He may well have used homemade stand oil in his 1940s Ripolin paintings, either as an addition to the paint or as a glazing medium to which he added his own pigments.[2]

By mid-1947 Nolan had completed more than forty-five paintings and numerous drawings on the Kelly subject. Twenty-seven paintings and nine drawings were selected for an exhibition, *The "Kelly" Paintings of Sidney Nolan, 1946–1947*, at the Velasquez Gallery in Melbourne from April 6 to 16, 1948. The paintings were arranged in a narrative sequence, with Kelly's spoken and written words, as well as extracts from the 1881 Police Commission report *Minutes of Evidence Taken Before the Royal Commission on the Police Force of Victoria*, contemporary newspaper articles, and J. J. Kenneally's *The Inner History of the Kelly Gang and Their Pursuers* (1929), used as extended titles (see figs. 30, 36, 37, 39, 41, 43). No photographs of the exhibition are known to exist, and it is not certain how the paintings were framed, although Warwick Reeder (1997) states that they were made from pale-colored tulip oak, a timber native to Australia. By 1956, he notes, these had been removed and replaced with thin edge strips.[3]

One of the paintings on exhibition at the Velasquez Gallery was given to a friend by Nolan prior to the show; the other twenty-six were retained by Sunday Reed, and she subsequently gave twenty-five of them to the National Gallery of Australia in 1977, where they were reunited with the previously gifted work. A lone painting from the original Velasquez group, *First-Class Marksman* (1946; fig. 30), remains separate, and is now in the collection of the Art Gallery of NSW. It was purchased at auction in 2010 for the highest price ever paid for a painting by an Australian artist.[4] Of the other Kelly paintings from 1945–47, most were returned by Sunday Reed to Nolan in 1969. A group of fifteen of these were subsequently given by Sidney and Mary Nolan to the Nolan Gallery at Lanyon. When the Lanyon gallery closed in 2007, the paintings moved to the Canberra Museum and Gallery. Other Kelly paintings were sold and dispersed, including two further works now in the Art Gallery of NSW, *The Camp* (1946; fig. 31) and *Robbed* (1947).

One of these dispersed works is Nolan's earliest painting on the subject, *Ned Kelly: "Nobody knows anything about my case but myself"* (1945; fig. 32). Given Nolan's own fugitive status, Kelly's words must have had a powerful personal meaning, and perhaps it is for this reason that the title is inscribed four times on the back of the cardboard support, once along each edge. In the painting, Kelly's face is visible through a slot placed low in the helmet but angled out of the horizontal, rather as if the subject is peering through an opening rather than being held within a three-dimensional helmet. The face seen through the slot is thinly painted, the color of the cardboard support providing the warm skin tones. In the paintings of Kelly that followed this early rendering there is little indication of a fully painted head under the helmet. In *First-Class Marksman* (see fig. 30), eyeballs are depicted floating in the opening against the landscape, as if there is no head at all beneath the mask, whereas in *The Camp* (see fig. 31) eyes emerge from a shaded interior, suggesting a dark emptiness within the helmet and imparting a powerful menace.

Unlike the other Kelly paintings from 1945–47, a head does in fact exist under the helmet of *Ned Kelly: "Nobody knows anything about my case but myself."* The portrait, later concealed by the black helmet overpaint, has been revealed by synchrotron-sourced X-ray fluorescence (XRF), and is imaged particularly well in the map of the element zinc, which locates the zinc oxide white pigment used in the Ripolin (fig. 33) (see sidebar on p. 42: Synchrotron-Sourced X-Ray Fluorescence). As he had previously done with *Boy in Township* in 1943, Nolan turned the painting upside down before applying a second working, in this instance the paint of the helmet, so the hidden head is now,

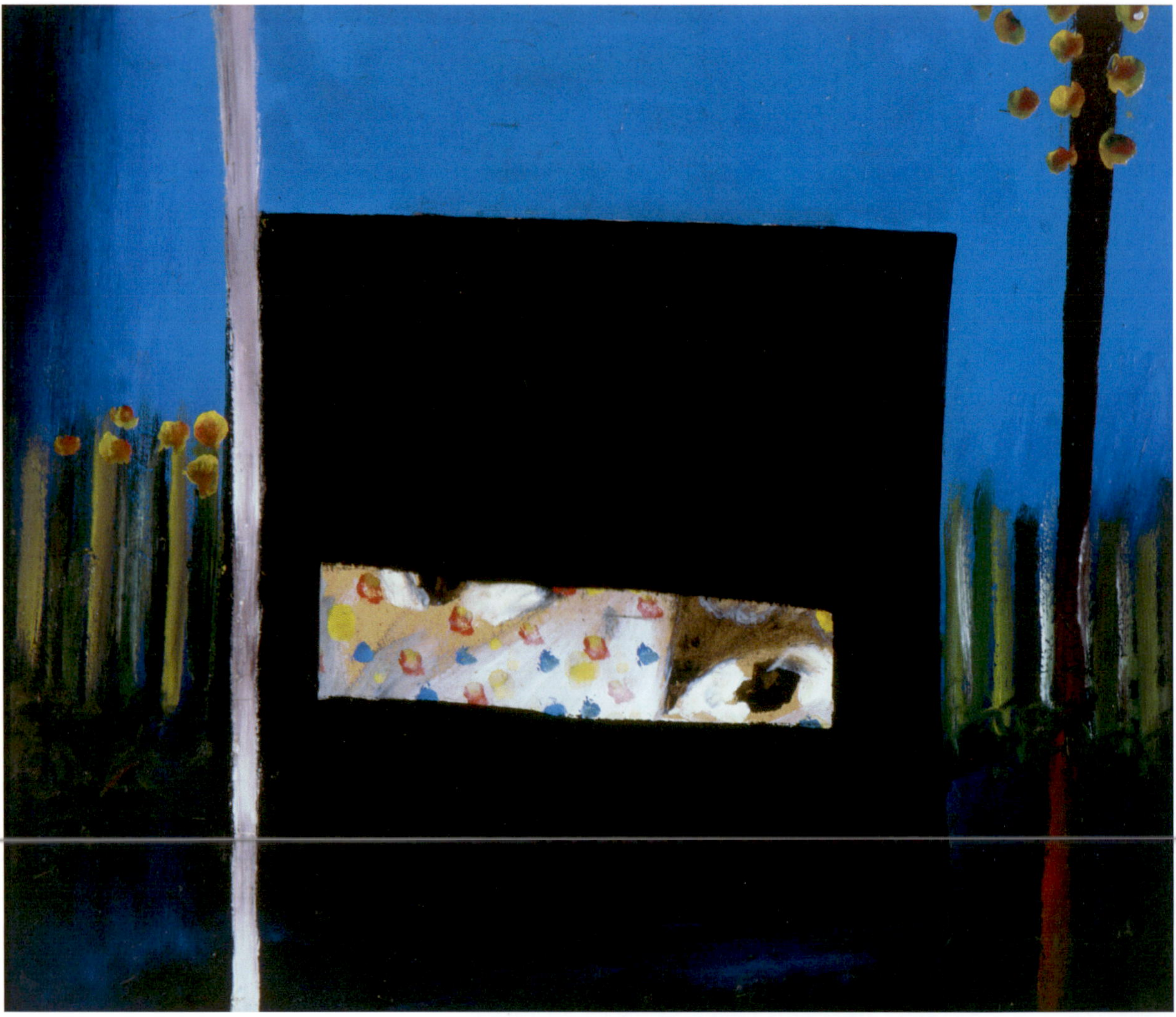

FIGURE 32
Sidney Nolan, *Ned Kelly: "Nobody knows anything about my case but myself,"* 1945. Ripolin oil enamel on cardboard, 63.0 × 75.0 cm (24¾ × 29½ in.). Heide Museum of Modern Art. Purchased with funds provided by the Friends of the Museum of Modern Art at Heide and Heide Circle of Donors 1998.

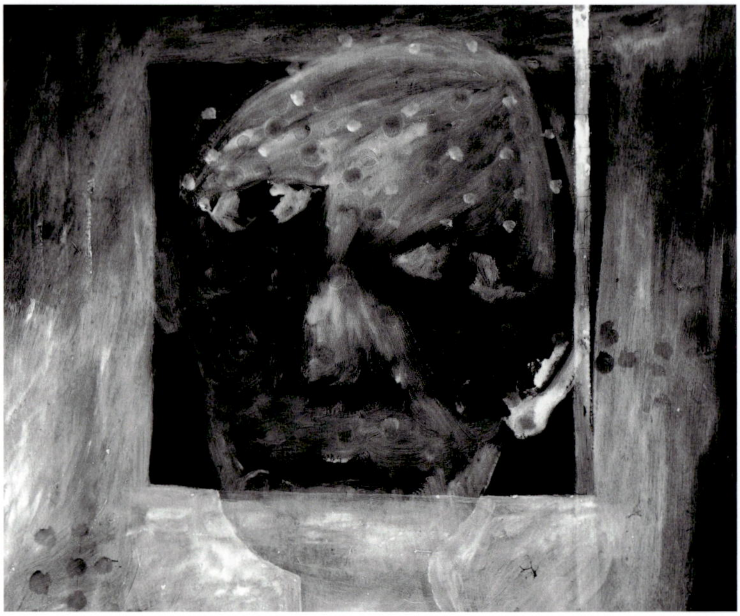

FIGURE 33
Zinc X-ray fluorescence map of *Ned Kelly: "Nobody knows anything about my case but myself"* (see fig. 32), rotated 180 degrees.

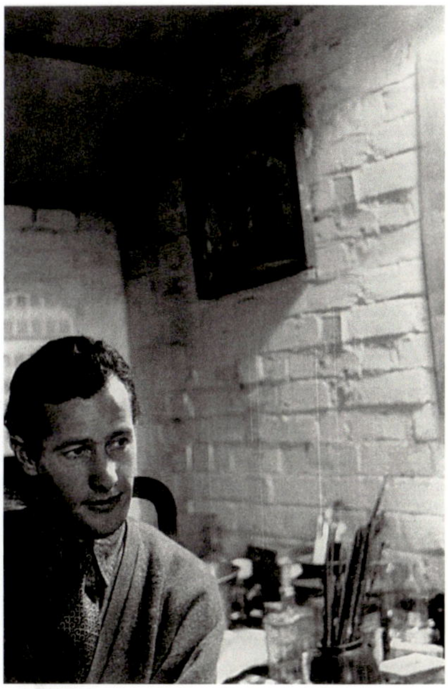

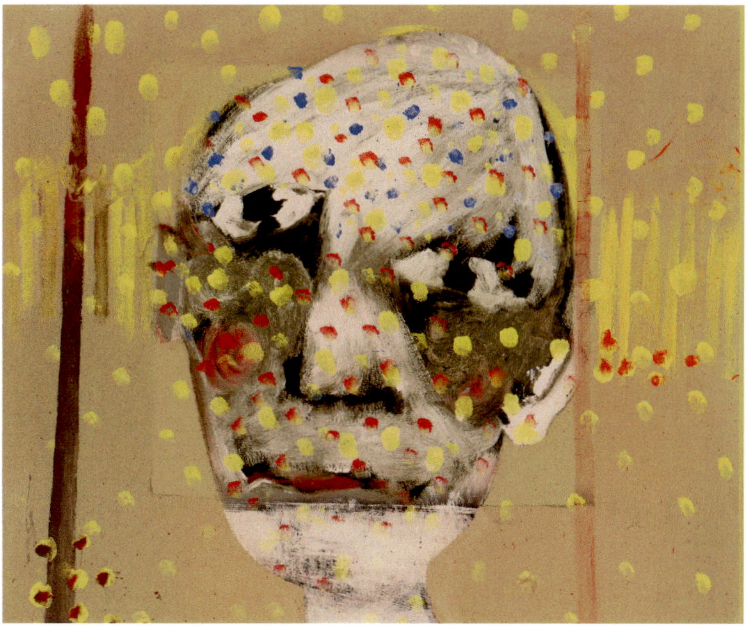

FIGURE 34
Albert Tucker, Nolan in Parkville studio, September 1944. Papers of Sidney Nolan, National Library of Australia, MS 10245, Series 2, File n. 151.

FIGURE 35
Color reconstruction of under painting revealed by X-ray fluorescence mapping of *Ned Kelly: "Nobody knows anything about mycase but myself"* (see fig. 32).

ironically, hanging by its neck from the top edge of the painting. The rotation of the hidden portrait accounts for the strange placement of the slot opening, more to the bottom of the mask than the top. The subject of the hidden portrait revealed by the XRF imaging is difficult to assign with certainty. Both Ned Kelly (as revealed by the death mask cast in plaster after his hanging) and Nolan (fig. 34) shared the physical features of a broad nose and rounded chin.

A digital color re-creation of the lower painting, based on the elemental signatures of four of the Ripolin paint colors Nolan used, shows a surprisingly complete portrait and a strange array of colored dots (fig. 35). Each color of dots is confined to a certain area of the painting: while the yellow dots (detected by the presence of chromium) are spread throughout, the red (barium) are restricted to the face (a group of red dots to the right side of the head are associated with the later reworking and introduction of flowering vegetation), and the blue (iron) are located only on the forehead. Whatever the intention of the dots, they imbue the portrait with a feeling of unease, whether of illness or disruption and anxiety. The whorl on the cheek, seen in both the red and white maps, is likely pink on the now hidden painting—the color re-creation being unable to replicate the color mixing of actual paints—achieved by applying red paint onto wet white paint and merging together with a finger. Interestingly, a cheek whorl or tattoo reappears decades later in Nolan's repertoire, in portraits he painted with spray cans, including a work titled *I Shall Tattoo Myself* (1982; see fig. 93), the title being a quote from one of Nolan's favorite poets, Rimbaud (Clark 2017). Perhaps, then, the portrait underneath the Kelly mask is a blend of all three figures: Kelly, Rimbaud, and Nolan himself. While this high-end imaging may have revealed a formative source for the conception of the Kelly figure in Nolan's paintings, *Ned Kelly: "Nobody knows anything about my case but myself"* continues to elude easy interpretation, retaining the mystery of its subject. Later, in 1984, Nolan hinted at the transference between the Kelly subject and himself: "Really the Kelly paintings are secretly about myself. You would be surprised if I told you. From 1945 to 1947 there were emotional and complicated events in my own life. It's an inner history of my own emotions, but I am not going to tell you about them" (Lynn 1985, 8–9).

Synchrotron-Sourced X-Ray Fluorescence

Synchrotrons are a source of superfast particles accelerated within a circular pathway with the use of magnets. These particles can be siphoned from the pathway into various beamlines with attached analytical instruments that study their interactions with materials. Synchrotron beamlines produce a flow of X-rays in a narrow and focused pathway that is more powerful than those produced by standard X-ray tubes in instruments used for art examination. The X-ray fluorescence microscopy beamline at the Australian Synchrotron, moreover, utilizes an X-ray detector, called the Maia detector, made from an array of individual detectors, all focused on measuring the fluorescence of particles or photons flowing back from the surface of a painting exposed to the X-rays. The combination of a focused beam and highly sensitive detection enables each 120-micrometer area of the painting to be measured in microseconds, allowing a complete elemental inventory of the painting to be built within twelve hours.

An abundance of photons at specific wavelengths can be assigned to a metal ion present in the paint, typically as a pigment giving a specific color. Thus, unlike standard X-radiography in which only heavy metals such as lead, often found in white artist's paints, are imaged, X-ray fluorescence measures a range of metals of increasing atomic weight, from calcium to gold. Combining the spectral results from each 120-micron area, elemental distribution maps of the painting can be built, each assigned to a metal ion present in pigments in the different paint colors.

The elemental assignments of each Ripolin color are known by the study of the pigments in the cans of paint from Nolan's Wahroonga studio. The elemental map of *Ned Kelly: "Nobody knows anything about my case but myself"* (see fig. 32) confirms that the white paint is pigmented with zinc oxide, the yellow with lead chromate, the blue with Prussian blue (iron ferrocyanide), and the red paint is made from an organic dye (PR 3) without a metallic assignment detected within the range of the Maia detector, but bulked up in the can with barites (barium sulphate), a semitransparent white used as a base for the dye. Each of the X-ray fluorescence maps for the elements zinc, chromium, iron, and barium can be assigned a color corresponding to the colors of the Ripolin paints; combining these four colors then gives an image of the full distribution of each throughout the painting, including those hidden below other paint layers, such as the paints used in the face that were later obscured by the Kelly helmet. The black carbon-based pigment in the paint of the helmet itself is not visible in these maps, as carbon is out of range of detection. A map of the element cobalt (not illustrated) shows the presence of a cobalt-based drier, used in greater (but still very minor) proportions in the black and dark blue Ripolin paints, which in turn corresponds with the helmet.

While X-ray fluorescence mapping is a powerful technique for analyzing paintings, it does have some limitations in distinguishing between early and late stages of a painting that utilizes a single set of paints. Some interpretation based on visual observation is therefore required to assess which parts of *Ned Kelly: "Nobody knows anything about my case but myself"* Nolan painted first, and which were confined to a second working session when the helmet was added.

FIGURE 36
Sidney Nolan, *Landscape: "The 'Kelly Country' is that portion of north-eastern Victoria which extends from Mansfield in the south to Yarrawonga in the north, and from Euroa in the south-east to Tallangatta in the north west,"* 1947. Ripolin oil enamel on Masonite, 121.4 × 90.7 cm (47⅞ × 35¾ in.). National Gallery of Australia. Gift of Sunday Reed 1977.

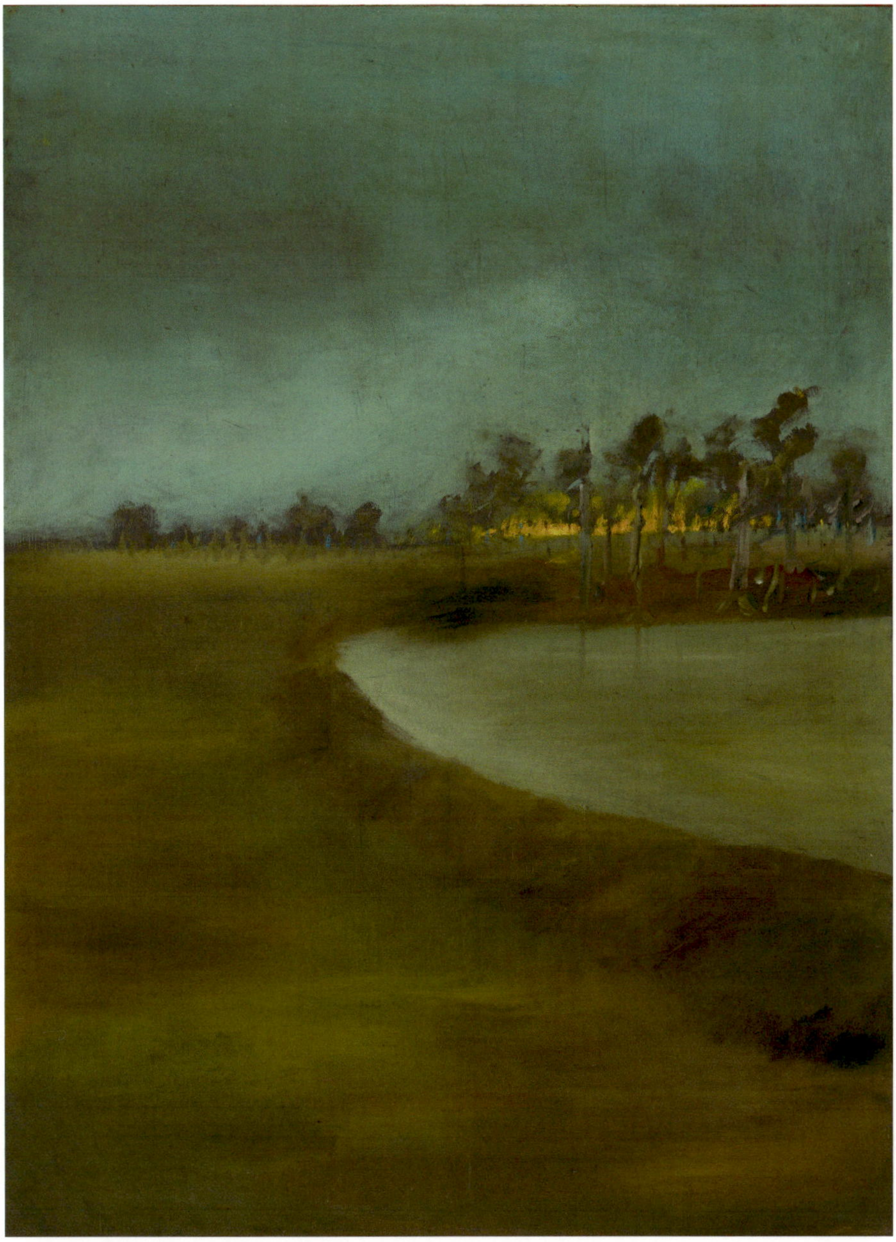

The earliest Kelly paintings, from 1945, were painted directly onto cardboard without a priming layer. Nolan began to use Masonite for the series in 1946, but the material remained difficult to obtain in the postwar period. This scarcity may explain why many of the Kellys on Masonite have paintings underneath the final versions—rejected ideas and compositions that Nolan overpainted, rather than discarding the valuable supports and working on a fresh board. Indications of these earlier paintings are visible texturally or through crack patterns (see fig. 44). The difficulty of sourcing Masonite may also have been the reason for one work, *The Evening* (1946–47), being painted on a sheet of plywood that already had a painting in place, *Ferris Wheel* (1945). Nolan simply turned the sheet over and painted the new image on the other side.[5]

With the exception of this single plywood painting, all the Kelly paintings dating after February 1946 are on Masonite boards, some with dark gray priming layers and others with a white horizontally brushed underlayer. According to Albert Tucker, Nolan began applying the colored Ripolin by laying in thin background washes on as many as six boards at once.[6] Landscapes were then fully worked up without figures using both matte and gloss Ripolin, which provided subtle variation in surface effects.

Although the narrative of the Kelly paintings is compelling, it is the rich coloring from the Ripolin paint and sense of changing light in the landscape that demonstrate Nolan's virtuosity with this difficult medium. The first work in the Kelly exhibition sequence at the Velasquez Gallery exhibition was in fact a poetic figureless painting, titled simply *Landscape* (1947; fig. 36). The light behind a line of trees, either a bush fire or a sunset, prefigures the storytelling of the series.

Nolan had begun to use a broader range of Ripolin colors on these works, expanding from his previously relatively limited palette of primary colors and green. He now mixed and blended colors both prior to application and directly on the boards, making for more modulation and subtly brushed boundaries. Figures are worked up thickly in gloss paint, with fine details of buttons, dress fabrics, and wallpaper patterns finessed at the last with a small brush. Even seemingly unintended effects produced by the Ripolin separating and pooling when painted over previously painted skies give subtle yet effective softened edges to tree lines (fig. 37, with detail fig. 38).

A 1963 catalogue of an exhibition of twenty-five of the Kelly paintings from the Velasquez Gallery exhibition noted that they had been recently cleaned, varnished, and reframed (cited in Reeder 1997, 9). The conservation work was undertaken by the artist Fred Williams. No documentation has been found to assist in determining the condition of the paintings prior to Williams's treatment or the extent of his intervention. The paintings had been extensively displayed from the 1940s to 1960s and endured less than ideal conditions, including exposure to cat urine, perhaps the cause of staining on *The Alarm* (1946; fig. 39, detail fig. 40), and dusting with an old shirt of John Reed's with the buttons still attached, which may have contributed to the scratches on many of the works (Harding and Morgan 2015, 280, 226). Williams's varnishes have now darkened and yellowed, and paint defects are visible, including blistering and crazing in some of the black paints, suggestive of heat or light damage (fig. 41, detail fig. 42); sharp-edged cracks revealing lower layers of paint, the upper paint layer apparently having been under physical stress (fig. 43, detail fig. 44); and a seedlike textural pattern in the light-colored areas on some works, possibly indicating the formation of crystalline zinc soap agglomerates (see sidebar on p. 71: Ripolin Zinc Soaps). The condition of these twenty-six paintings is in contrast with *First-Class Marksman* (see fig. 30), which Sunday Reed separated from the Velasquez Gallery group after the works had been exhibited and toured to Paris in 1949. *First-Class Marksman* is in exceptional and original condition. Never having been varnished, it retains the surface contrasts inherent in the gloss and matte Ripolin paints, and there are no indications of any of the deterioration features observed in the other works in the group. This suggests that the condition issues observed in the National Gallery group are related to the conditions that the works experienced after 1949.

Observations of the condition of Nolan's paintings in Ripolin are important when considering the viability of commercial paints used as artist's paints. House paints are generally formulated for a maximum of ten years' exposure to harsh outdoor light and humidity. Artist's paints are expected to have longer stability but are unlikely to be exposed to the extreme environmental conditions of household paints. Questions regarding the vulnerability of Nolan's Ripolin paintings to the effects of aging can only be addressed through a full understanding of the paint's composition. We are assisted in this endeavor by having access to Nolan's paint collection, which remained in his

FIGURE 37

Sidney Nolan, *Burning at Glenrowen: "I got no answer, of course, and I looked in and found the bodies of Dan Kelly and Steve Hart lying together. As far as I could tell they were burnt from the waist up," Very Rev. Dean Gibney*, 1946. Ripolin oil enamel on Masonite, 121.5 × 90.7 cm (47⅞ × 35¾ in.). National Gallery of Australia. Gift of Sunday Reed 1977.

FIGURE 38

Detail of Ripolin paint separation and pooling on *Burning at Glenrowen* (see fig. 37).

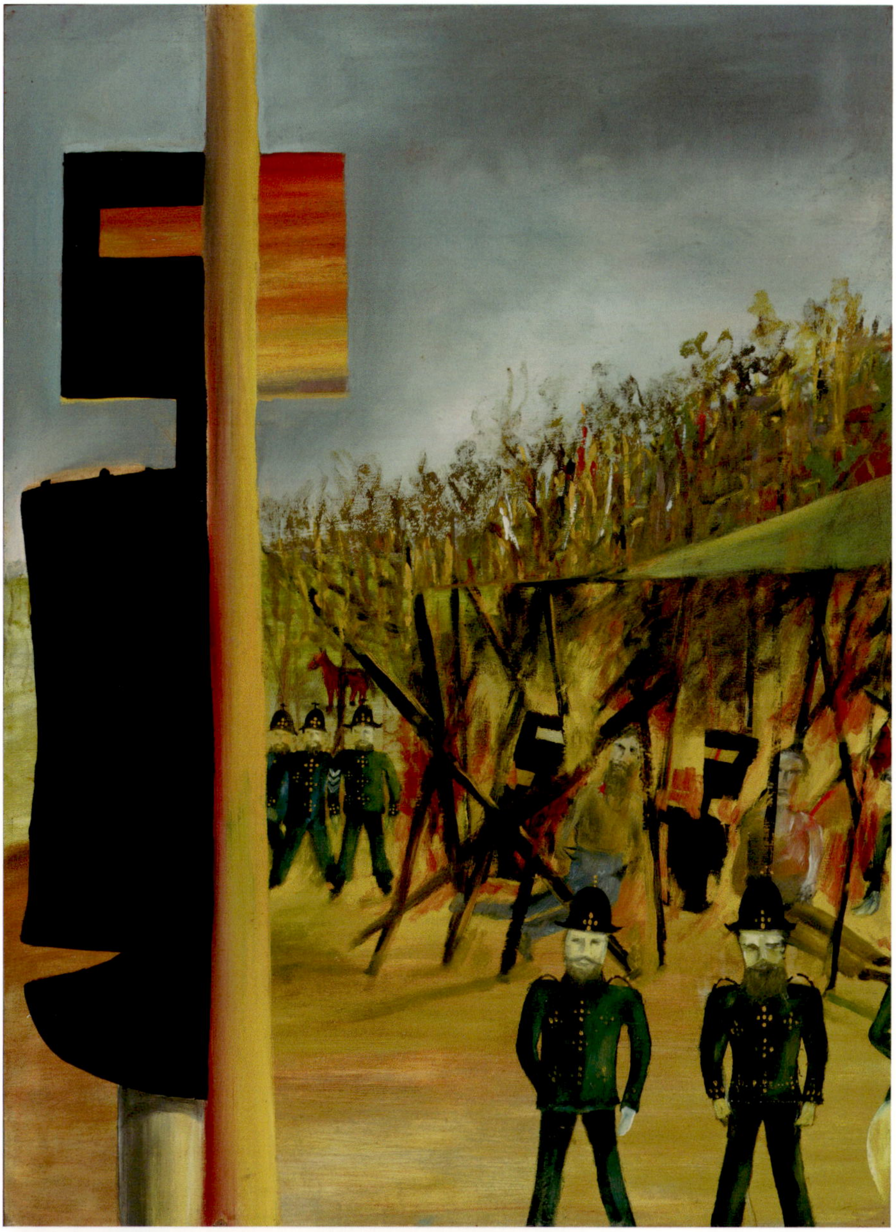

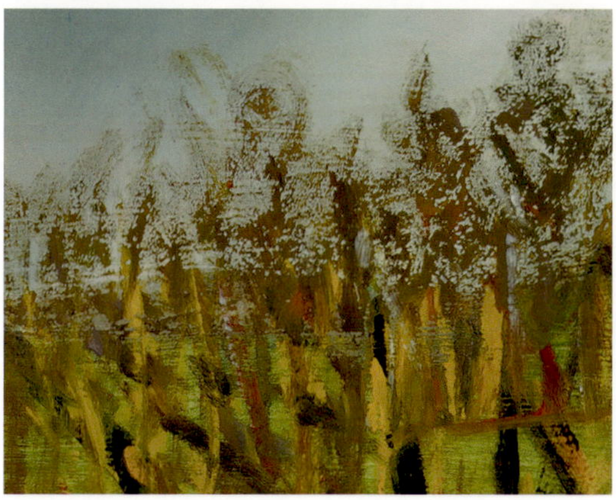

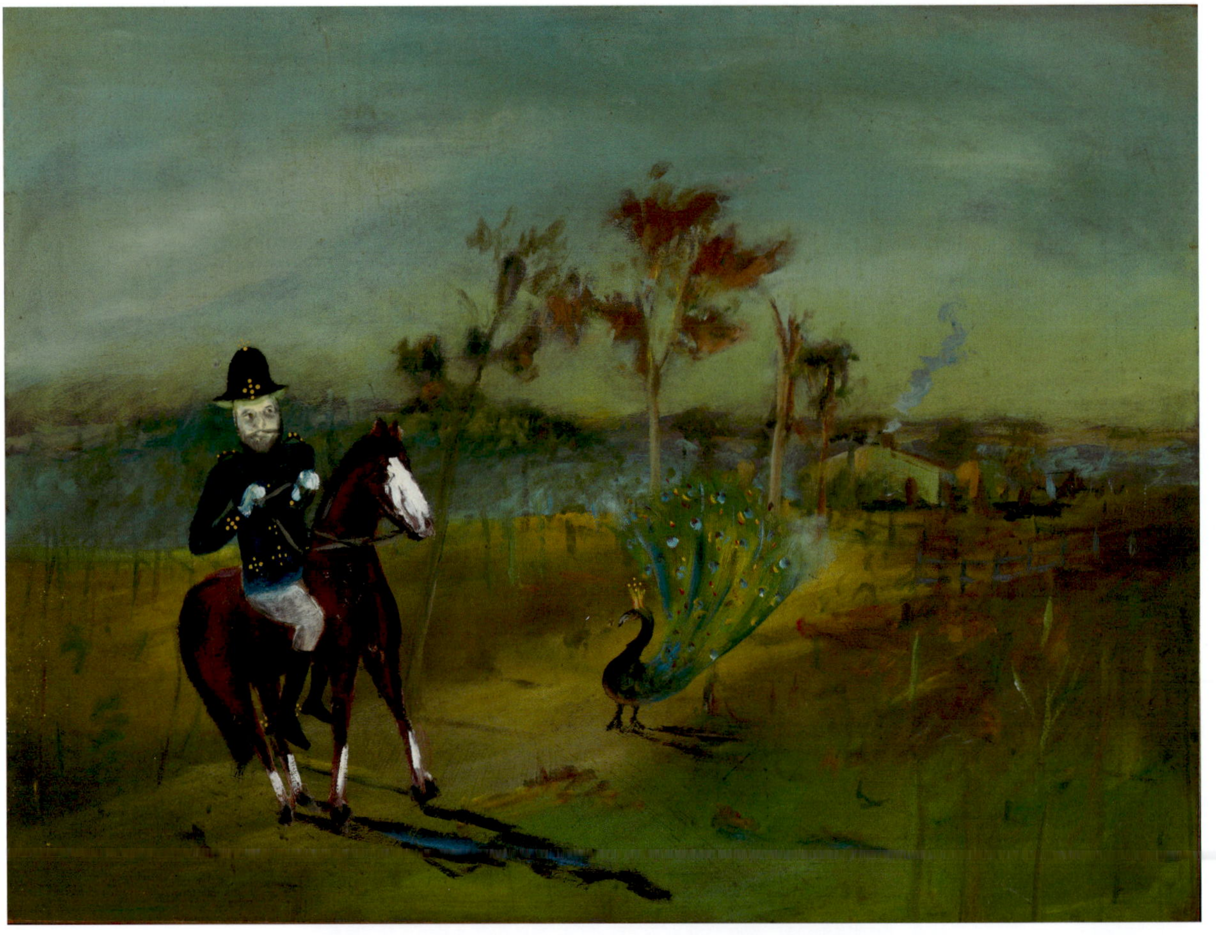

FIGURE 39
Sidney Nolan, *The Alarm: "A peacock which used to live on the top of a tin shed. The bird was accustomed to making warning cries at the approach of any stranger,"* 1946. Ripolin oil enamel on Masonite, 90.5 × 121.3 cm (35⅝ × 47¾ in.). National Gallery of Australia. Gift of Sunday Reed 1977.

FIGURE 40
Detail of stains, likely cat urine, on *The Alarm* (see fig. 39).

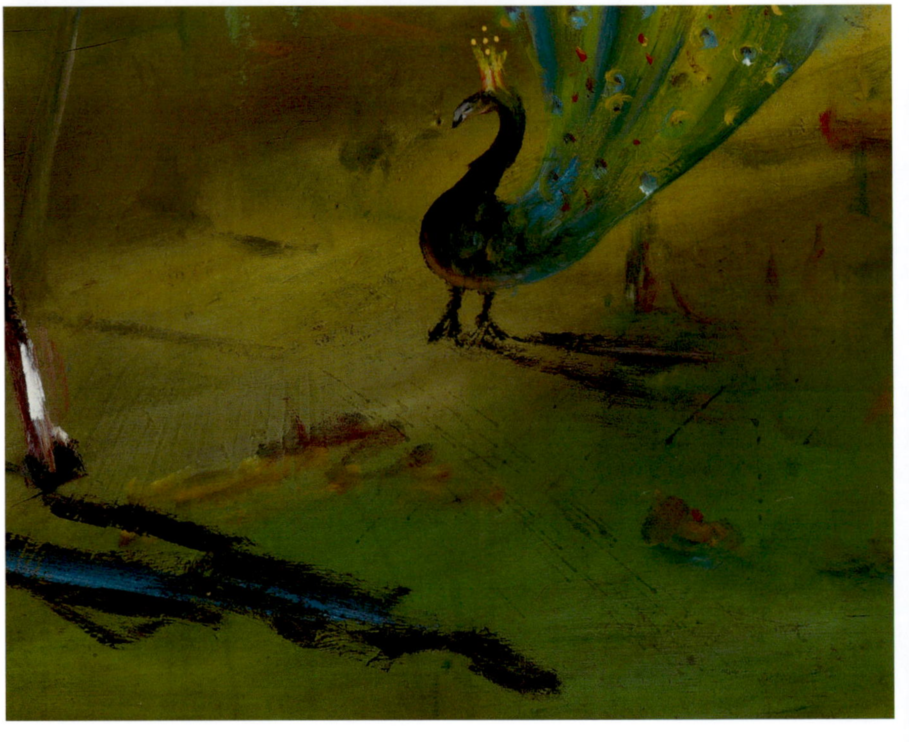

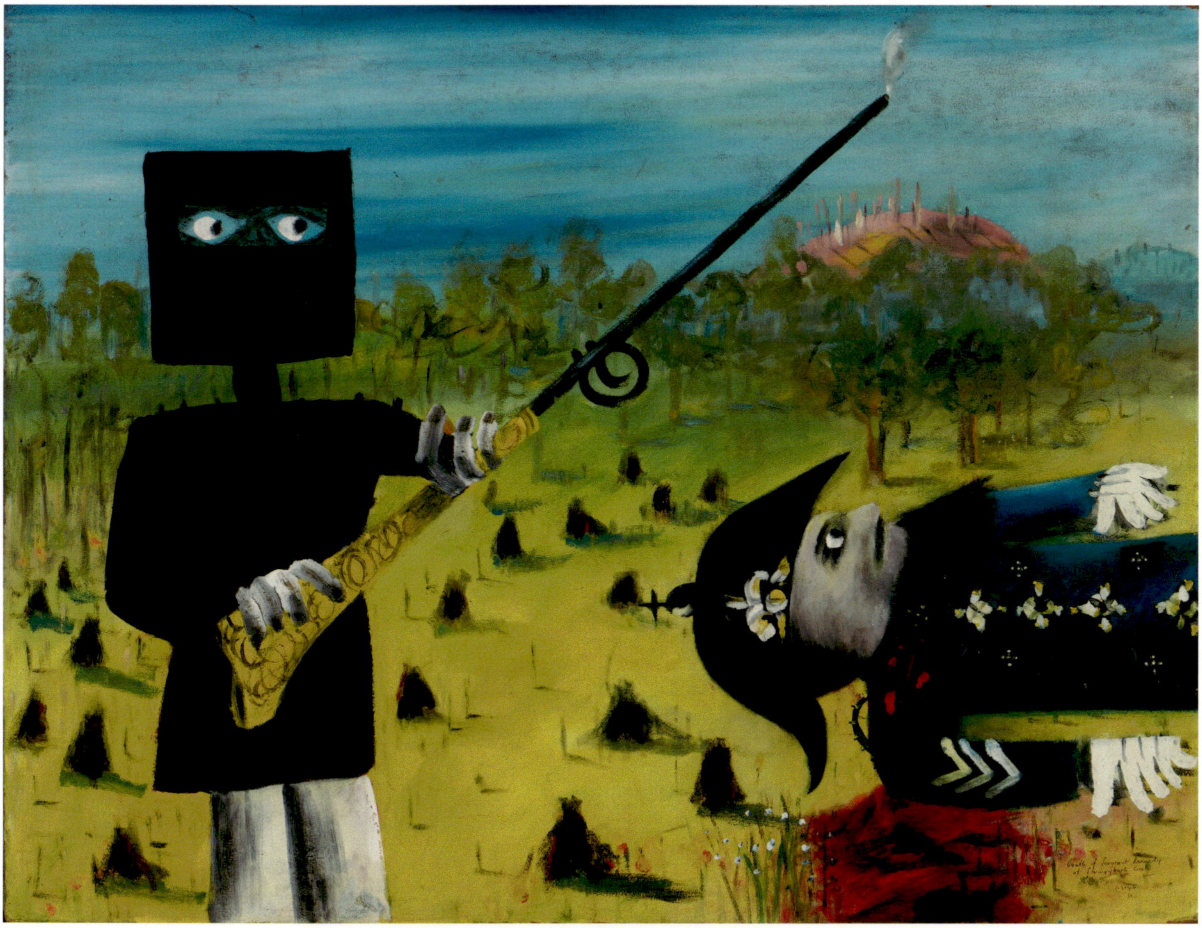

FIGURE 41
Sidney Nolan, *Death of Sergeant Kennedy at Stringybark Creek: "Kelly: 'Had he been my own brother I could not help shooting. I put his cloak over him and left him as honourable as I could,'"* 1946. Ripolin oil enamel on Masonite, 91.0 × 121.7 cm (35⅞ × 48 in.). National Gallery of Australia. Gift of Sunday Reed 1977.

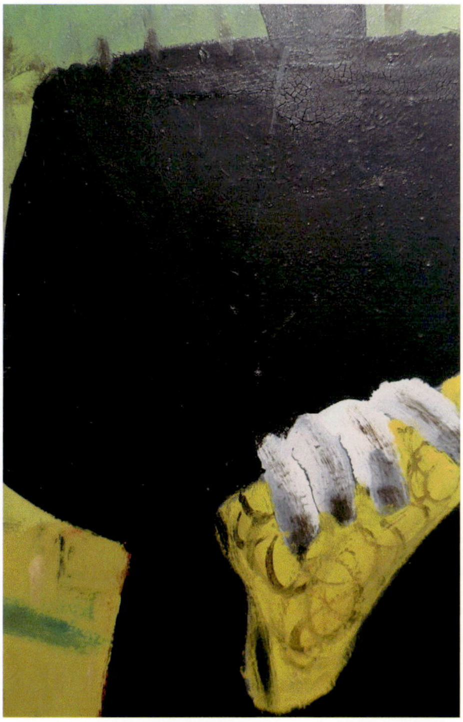

FIGURE 42
Detail of paint crazing on *Death of Sergeant Kennedy at Stringybark Creek* (see fig. 41).

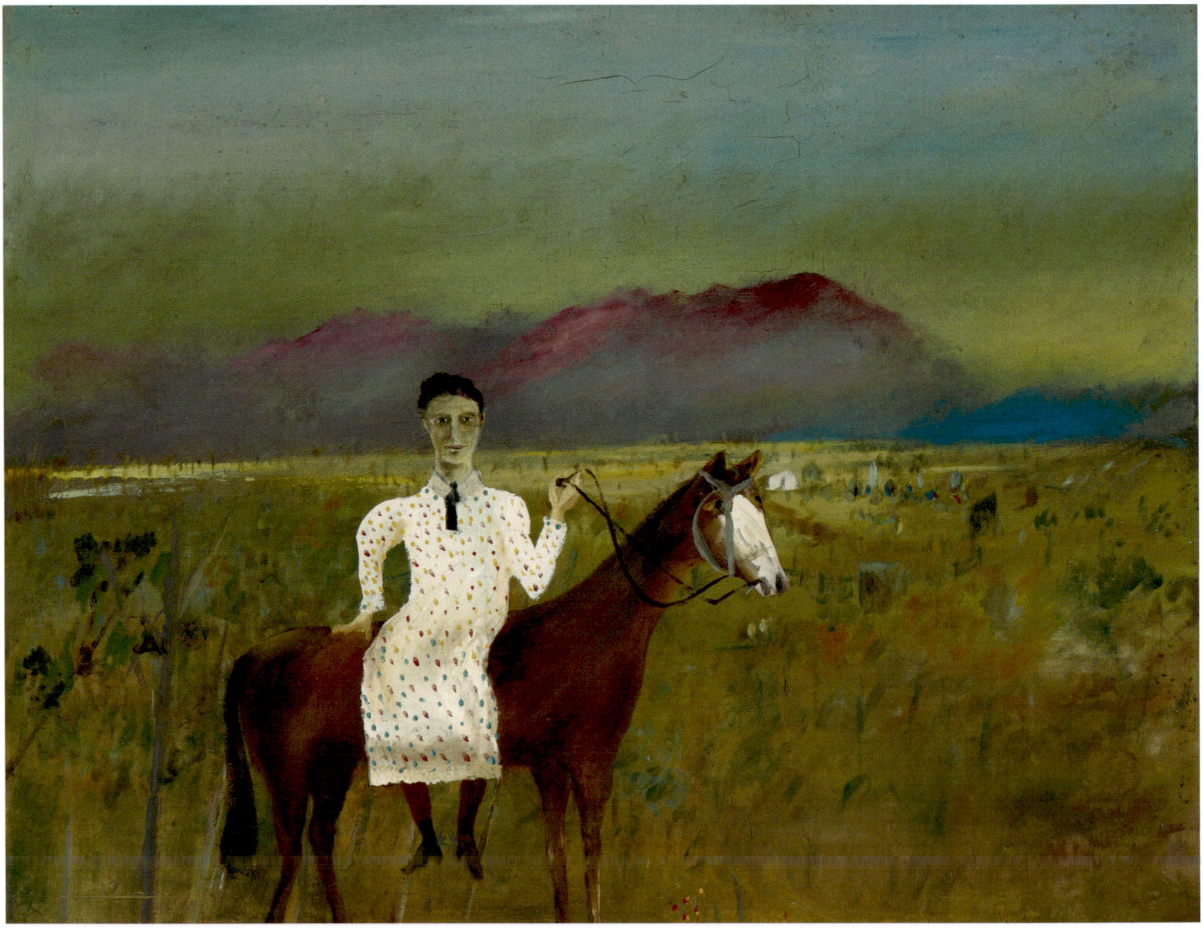

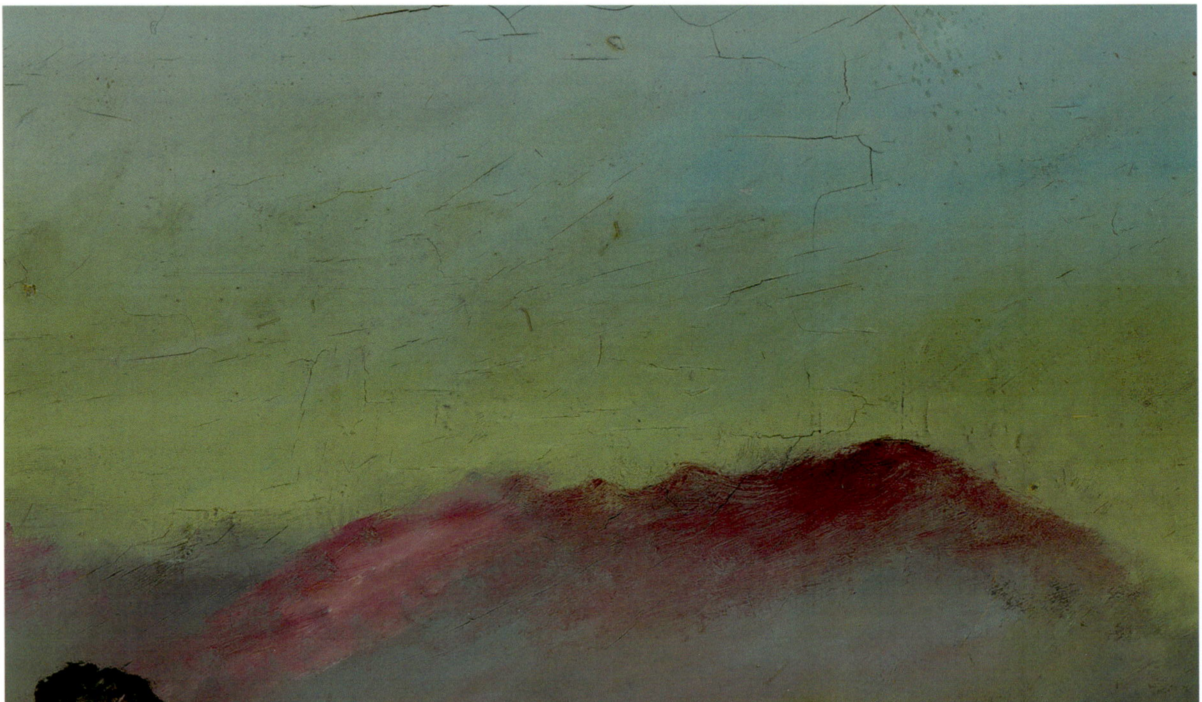

FIGURE 43 (top left)
Sidney Nolan, *Steve Hart Dressed as a Girl: "He appears to have been possessed of a considerable courage and resource, and during the period of his outlawry frequently rode about in feminine attire,"* 1947. Ripolin oil enamel on Masonite, 90.6 × 121.6 cm (35⅝ × 48 in.). National Gallery of Australia. Gift of Sunday Reed 1977.

FIGURE 44 (bottom left)
Detail of linear cracks in sky exposing lower painting in black on *Steve Hart Dressed as a Girl* (see fig. 43).

studio in Wahroonga, a suburb of Sydney, and was gifted by his daughter Jinx Nolan to the Art Gallery of NSW in 2007 (fig. 45).[7]

Ripolin

In 1953, Nolan packed up and moved to London. The studio in Sydney, which he left behind more or less untouched, survived as a time capsule of his materials. Unlike most studios preserved at the time of the artists' death, such as those of Francis Bacon (1909–1992) and Jackson Pollock, Nolan's Wahroonga studio is unique in representing a mid-career moment in the artist's long and prolific working life. Although the materials are no longer retained in the context of the studio itself, they remain a valuable resource for understanding the artist's engagement with house paints, solvents, pigments, and the wartime reuse of containers. Among the materials are forty-seven cans of prewar Ripolin paint, a medium not only critical for the study of Nolan's processes and techniques but of a type used by other twentieth-century artists, including Pablo Picasso, Wassily Kandinsky (1866–1944), and Francis Picabia (1879–1953) (Casadio et al. 2013; King et al. 2013). Although it is not certain that the Wahroonga cans of Ripolin were the very paints that Nolan used on his paintings in Melbourne from 1942 to 1947, including the seminal Kelly paintings, they are of a prewar type and may have been brought into the studio by Nolan from an earlier source.[7]

Ripolin oil enamel was first manufactured in Holland in the 1880s, developed by a merchant, Gustav Briegleb, and a paint chemist, Dr. Carl Julius Ferdinand Riep (Raeburn 2011). The manufacturing process and ingredients were a closely held secret. Ripolin paint became renowned in the early twentieth century for the intensity, range, and consistency of its colors, as well as its flexibility and good adhesion to all types of supports. It was promoted as a car paint, as a highly reflective interior paint during wartime blackout conditions, and even as a germicidal finish for hospitals due to its ability to flow and fill small crevices, producing smooth, easily cleaned surfaces.

The name Ripolin appears to be a blend of *Riep* (the developer's name) and *linseed* (presumably after the oil used in its formulation) (Raeburn 2011). In 1897, a French

FIGURE 45
Contents of Nolan's Wahroonga studio, Artists' Materials Archive, Art Gallery of NSW.

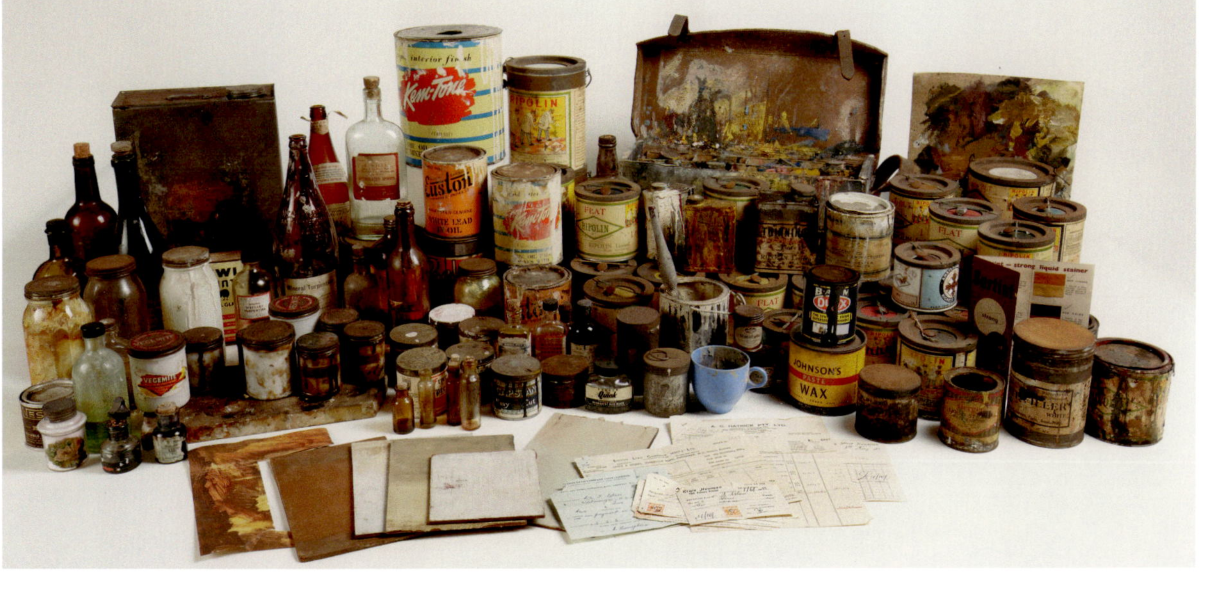

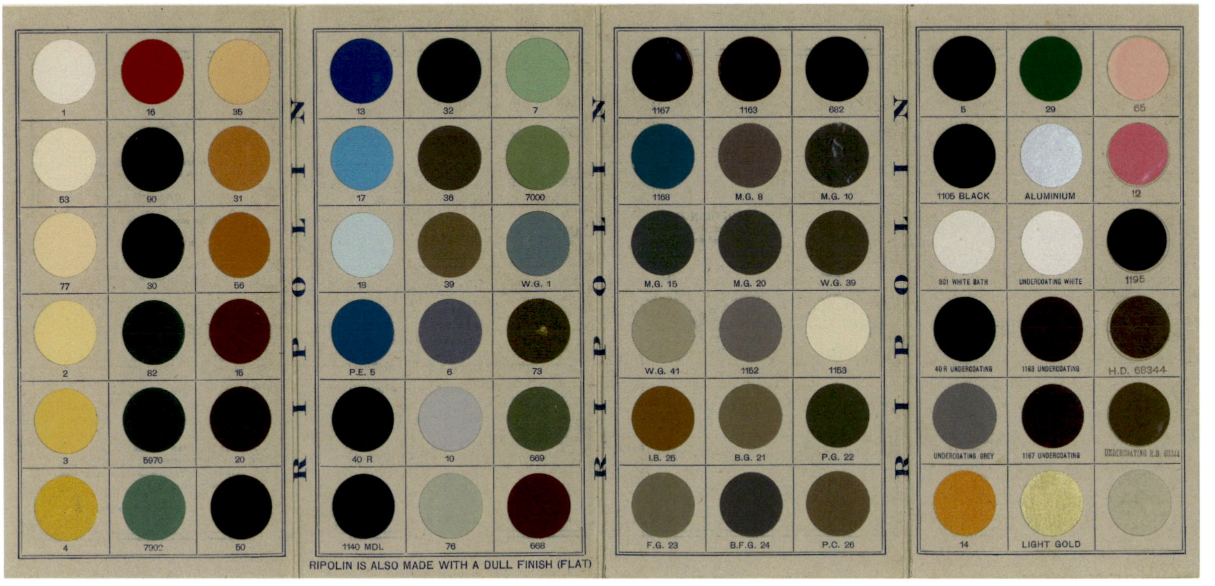

FIGURE 46
Ripolin Ltd., *Paint Colour Chart* (inside), c. 1923. Walter Buxton Bruce Collection, State Library of South Australia.

branch of the company, Le Ripolin, was formed in partnership with Lefranc, an artist's paint manufacturer, and began manufacturing in France. The following year, the Ripolin Ltd. Company was established in England, initially importing and selling the paint manufactured at the Dutch factory ("Ripolin Celebrates Its 80th Anniversary" 1978). It was the English-labeled—but Dutch-manufactured—paint that was exported to Australia from the 1910s into the 1920s.

Although advertisements for Ripolin paint did not appear in Australian newspapers until the late 1910s, it must have been available in Sydney as early as 1914, when the Australian distributor promoted its use in the new Nepean Hospital ("Our Hospital" 1914). Although at first it was available only in white, in 1921 the full color range arrived in Australia, attracting great interest ("Crane's 'the Paint Specialists'" 1921). A chart with seventy-one color swatches dating from the 1920s offers the English numbering of the Ripolin colors (fig. 46). The French Ripolin colors were similarly numbered, as documented on an interwar chart of Le Ripolin dated 1929–46, suggesting a reciprocity between the French and English Ripolin ranges, even if they were manufactured in different factories (Casadio and Gautier 2011, 138–39).

Because Ripolin originated in Holland, it was initially subject to high import tariffs and therefore expensive in Australia compared to locally manufactured products. In 1932, a Ripolin factory was opened in Southall, London;[8] this now became the source of Ripolin in Australia, bringing reduced Commonwealth tariffs. The lower price must have improved its marketability, but this advantage was short-lived as the outbreak of war in 1939 caused the cessation of all domestic paint imports from Britain.

The intensity and variety of the prewar Ripolin color range was a distinct advantage compared to the limited colors available in locally produced paints, and no longer confined Nolan to the three Dulux primaries he had been using in 1942. The cans left in Nolan's studio in Wahroonga give an indication of his color preferences (for a full list of colors in the collection of the Art Gallery of NSW, see appendix II). In addition to their brilliancy, some colors were formulated without any opaque white pigment, giving a semitransparent effect. These were unlike typical house paints, in which the coverage provided by the white pigment was more valued.

Nolan's Wahroonga Ripolin is characterized by two types of finish: flat (eleven different tints) and gloss (seven tints). Most of the gloss cans are described on their labels as "Quality BLT." This term appears on the English-labeled gloss Ripolin from at least 1927, but the meaning of the abbreviation remains elusive.

The Ripolin paint in Nolan's Wahroonga studio was all made at the London factory, according to the labels on the cans, and must therefore postdate 1932. Although Nolan contended that he had obtained the Ripolin in the early 1940s directly from London (B. Smith 1962), this seems improbable given that imports of nonessential civilian materials were suspended during the war. More likely, Nolan and Sunday Reed obtained the paint from Ripolin House in Sydney—as suggested in Nolan's letter to Sunday of December 1942—which had a good supply just prior to the war (*DPA&NZ,* March 10, 1939).

In 2009, an analytical study of the Wahroonga studio Ripolin paints commenced. Prior to this study, it had been unclear what Nolan's Ripolin paint was made from, with descriptions varying between enamel, synthetic, alkyd, and oil. The features of the paint had also only been studied in terms of dry surfaces and effects on paintings, rather than the working properties of the wet paint. By painting out each color, we revealed the extraordinary quality of these paints after sixty years in the can. The gloss colors that are without white solids, such as Blue Deep 40RD, Black 1105, and Red 16, were all still well dispersed with virtually no pigment settlement, whereas those containing zinc white had settled, a still-pliable lump of solids resting in the bottom of the can with the liquids above (fig. 47). In most cases the gloss paints flowed out to form perfectly smooth, mirrorlike surfaces. Tinting strength of all the colors was very high. In the case of both the gloss and matte Blue Deep 40RD, the paint-outs were so saturated with pigment that the dried films were iridescent. Notably, one gloss color of the Ripolin paint-outs—Black 1105—showed a tendency to wrinkle on drying; this occurred in the surface skin of the paint within several days of painting out, though after two weeks the wrinkling stabilized and did not develop further.[9] There is evidence of wrinkling in the glossy black paints on many of the Kelly paintings, including *First-Class Marksman*

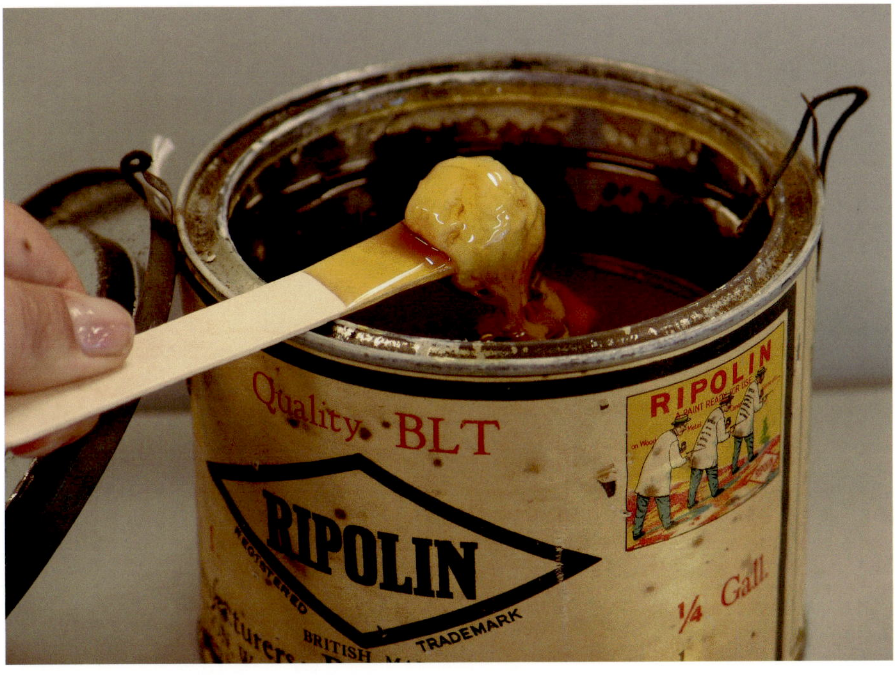

FIGURE 47
Ripolin BLT White No. 1, lifting solids from liquids. Sidney Nolan Wahroonga studio paints. Artists' Materials Archive, Art Gallery of NSW.

(see fig. 30) and *The Camp* (see fig. 31), and it is likely that Nolan observed this occurring as the paint dried over the first few days.

Another physical feature of a number of the paint-outs was the presence of small, hard aggregates that formed seedlike lumps. These lumps were identified with FTIR spectroscopy (see sidebar on p. 17: FTIR and XRF Analysis) as crystalline zinc carboxylates such as zinc stearates and palmitates, often found in oil-based paints made with zinc oxide. Additionally, one paint, Pink Very Light 565, was found to have a brittle waxy layer at the interface between the solids and the liquids when the solids were lifted from the can. This, too, was identified as zinc stearates/palmitates, but in this instance, given its different physical distribution, they may have been added to the paint as a stabilizer rather than formed in situ (van den Berg, Burnstock, and Schilling 2019).

Zinc carboxylates are associated with unstable aging of paint films, often developing seedlike aggregates and migrating to the surface, breaking the paint film as they emerge many years after the paint has dried and hardened (see sidebar on p. 71). In the case of Ripolin paints on artworks by Nolan, they do not generally appear to have developed into this more problematic crystalline form but remain well distributed within the network of the paint matrix, unlike the lumps that we see in the paint-outs from the cans themselves. The difference between the dry Ripolin paint on Nolan's paintings and the Ripolin paint from the cans suggests that the long exposure of the zinc white pigment to unoxidized fatty acids in the oil of the liquid paints has allowed these stearate/palmitates to form, whereas on the paintings the oils dried and oxidized quickly.

Scientific analysis of the collection of Nolan's Wahroonga paint cans confirmed the complex formulation of the British-made Ripolin. FTIR spectroscopy of the liquids identified the principal binder as oil. An absorbance peak at 968 cm^{-1} in samples of the still-wet paint is associated with C-H out-of-plane deformation *trans* double bonds, indicating stand oil, a pre-polymerized oil (Meilunas, Bentsen, and Steinberg 1990).[10] This absorbance peak rapidly diminishes on the drying of the paint, making it impossible to distinguish stand oil using FTIR on samples taken from paintings. Meth-Prep gas chromatography–mass spectroscopy (Meth-Prep GC-MS) analysis undertaken at the Getty Conservation Institute on samples decanted from the cans provided further detail on the types of oil, additions, and the pre-processing of the binders (Dredge et al. 2013). Meth-Prep GC-MS confirmed the presence of drying oils in all samples of Ripolin, as well as copal resin and colophony in some of the paints (see appendix II, table 4). The gloss paints contained a higher percentage of fatty acids (2.7–4.9%) from the oil, compared with the flat paints tested (0.9–1.3%); the analytical results were supported by visual observation of the paint in the cans: the liquids in the gloss paints were thick and oily, while the flat-paint liquids were thin and fluid, dominated by the presence of volatile solvents.

The precise identification of drying oil present in paint is a difficult analytical challenge. Determining the ratios of palmitic to stearic fatty acids (P/S) by GC-MS analysis can indicate oil type. However, this methodology is problematic with regard to twentieth-century oil paints due to the increased range of oil types used, many of which, such as tung and castor oil, exhibit P/S ratios similar to that of linseed oil, in the region of 1.5 (Schilling, Mazurek, and Learner 2007, 130). The P/S ratio identified for Nolan's cans of Ripolin paint was 1.3–1.5, leaving us unable to say whether Ripolin paints contained one of these three common oils, or perhaps even a mix. Fatty acids found in specific types of oil, called fatty acid markers, are present in a small number of oil types, including castor oil, whose marker, ricinoleic acid (12-hydroxy-9-*cis*-octadecenoic acid), was identified in the mass spectra of the Ripolin White No. 1 paint. Although a slow drier, dehydrated castor is a pale oil compared to linseed and so would have been a good choice for the white paint color. Because castor oil also forms a more flexible paint film, it may have been used in the white color that was pigmented with zinc oxide, an inherently brittle pigment. The same fatty acid marker for castor oil was not identified in the other colors, however, suggesting that this oil type was exclusive to White No. 1. The oil in the other cans could not be firmly attributed to either tung or linseed, since the tung oil marker, α-eleostearic acid (*cis, trans, trans*-9, 11, 13-octadecatrienoic acid), rapidly oxidizes (Schönemann et al. 2006), while linseed does not have a distinctive marker.

GC-MS analysis of Nolan's Ripolin paints confirmed the FTIR analysis regarding the presence of stand oil. This ingredient is indicated by the proportion of suberic to azelaic fatty acids (Sub/A) above 0.4 in all but one of the paints tested (van den Berg, van den Berg, and Boon 1999), the single exception being the flat Black 505 containing bone black, with a Sub/A proportion of 0.3 (see appendix II, table 4).

Copal, a fossilized tree resin, was identified in several of the Ripolin samples using GC-MS by the marker eperuic acid (Dredge et al. 2013). The eperuic acid was well oxidized, the hard, insoluble resin probably having been heated to a high temperature to liquefy it before "running" it into hot oil. The addition of copal resin to the Ripolin paints would have imparted gloss and hardness and assisted with drying, although it could also increase the tendency of the paint to yellow and become brittle (Standeven 2011, 36). Colophony is even less stable than copal and is therefore a surprising addition to the Ripolin paints, which were reputably of the highest quality. It is present in a number of color tints along with copal; however, in other colors colophony is the only resin detected, and so cannot be considered simply as an adulterant of the copal. The abietic acid colophony marker, identified with Meth-Prep GC-MS, was so oxidized that it had completely converted to 7-oxo dehydroabietic acid (Dredge et al. 2013). This suggests significant heat treatment of the colophony as a form of processing.

Processed colophony products, more stable than unprocessed colophony, include ester gum, a glossy resin, and driers produced by reacting colophony with metallic ions (Standeven 2011, 29–30). Although metallic driers are in many of the Ripolin paints—detected with X-Ray Fluorescence (XRF) by the presence of minor amounts of the elements cobalt and manganese—it is unknown if they are resinate driers made with colophony or naphthenates. Therefore, the exact form in which the colophony was added to the Ripolin paints has not yet been identified. However, the addition of these types of "cooked" materials—heat-bodied oils, processed colophony, and liquefied copal—to the Ripolin paints is consistent with the historically documented types of processing common to the preparation of oil enamels of the period (Standeven 2011).

The white and pale colors in the Ripolin paints are dominated by the pigment zinc white (zinc oxide). Compared to lead white, zinc white is a cold-hued white, more brittle and more expensive. The small amount of chrome yellow (lead chromate) detected in the white Ripolin paints was probably added to warm the tint, but it would also have helped the paint to dry. Zinc white is also less toxic than lead-based paints. While France had banned the use of lead in household paints as early as the 1920s, Australia continued its use until after the Second World War, when titanium white (titanium dioxide) was produced in sufficient quantity at a factory in Tasmania. The analytical detection of zinc-based enamel paints on works by Australian artists is therefore suggestive of an imported paint product such as Ripolin.

Prussian blue (iron ferrocyanide) is the principal blue in several of the Nolan Ripolin colors, either as a deep transparent blue (gloss 40RD and flat 540RD) or mixed with zinc white as a pale blue (PE5). The transparency and richness of the deep blue tint would have been extraordinarily appealing to Nolan. Ultramarine 13, despite its name on the can label, is pigmented not with French ultramarine pigment but with phthalocyanine blue (copper phthalocyanine), a pigment not produced until 1935 (Gettens and Stout 1966). If all the Wahroonga Ripolin cans are contemporary with one another, the presence of phthalocyanine blue in this hue indicates that the paints were manufactured after 1935. The flat green paints Peacock 582 and Green Dark Yellowish 590 both combine Prussian blue and chrome yellow (lead chromate) with zinc white, whereas Olive Green 573 is a mixture of carbon black and chrome yellow. Flat Lemon Yellow 514 and flat Canary 504 are both based on chrome yellow pigment, but the Canary is modified by the addition of lead sulphate. Yellow Ochre 3056 is a mixture of yellow ochre (iron oxide) and chrome yellow. The Red 16 (gloss) and 516 (flat) contain the coal tar pigment toluidine red PR3 (o-nitro-p-toluene-azo-β-napthol, $C_{17}H_{13}N_3O_3$) and barites (barium sulphate). The brilliant midtone of toluidine red led to its common use as the official government red color through the twentieth century (on post boxes for example.) The flat Black 505 contains bone black, a pigment made from charred animal bones. Bone black was not detected in the gloss Black 1105 can, however, which is pigmented instead by a carbon black. Most of the flat colors contain, in addition to colored pigments, kaolin (aluminum silicate) and talc (magnesium silicate), which would have assisted in the formation of matte surfaces on drying. The pigments detected in the cans of Ripolin are of only a few different types, but when used in a variety of combinations they give a large range of colors. Representing the best of the available powerful tinting pigments of the period, they could be used in small quantities to impart strong color while leaving the paint fluid.

An Anomaly

First-Class Marksman (see fig. 30) is typical in technique of the Kelly paintings. The only one of the Velasquez Gallery exhibition works not painted at Heide,[11] it shares with the other Kelly paintings the deliberate and distinctive use of gloss and matte passages of Ripolin paint. Unlike the others of the group, this painting was not varnished, and these critical surface variations are well preserved. Nolan primed the smooth side of the Masonite sheet with a matte white paint, which remains visible between colored brushstrokes or through thin and semitransparent paints. He then worked the bushland in the middle ground in the flat Ripolin paint, with final highlights on leaves and branches in gloss. The figure, hills, and sky are all more densely worked in the gloss medium. The figure of Kelly himself is rendered in high-gloss, opaque black paint. The edges of this figure are sharply delineated and sit separate from the paint layers of the landscape, as if the black paint of the Kelly armor was applied after the rest of the painting had dried, unlike the black paint of his fingernails, which swirls and merges with the wet paint of the fingers. When the black paint on the Kelly armor of both this painting and another Kelly from 1946, *The Camp* (see fig. 31), were analyzed with FTIR spectroscopy, alkyd resin was identified as the paint binder. This unexpected result indicated that the black paint of the Kelly armor on both these works was not Ripolin.

A sample of the black paint on *The Camp*'s Kelly armor, shown in cross section (fig. 48 top), reveals that the armor has been painted twice. When the cross-section sample is examined microscopically under ultraviolet fluorescent light (see fig. 48 bottom), the earlier, oil-based paint (Ripolin) has a distinctively brighter appearance than the alkyd paint applied on top. A similar result was obtained for *First-Class Marksman*, suggesting that the reworking of the black figure was not a sole example: at least two, if not all, of the Kellys received this repaint.[12] Visual examination of the Kelly armor on both

FIGURE 48
Microscopic sample of black paint from Kelly armor of *The Camp* (see fig. 31), mounted as cross section: incident light (top) and ultraviolet light (bottom).

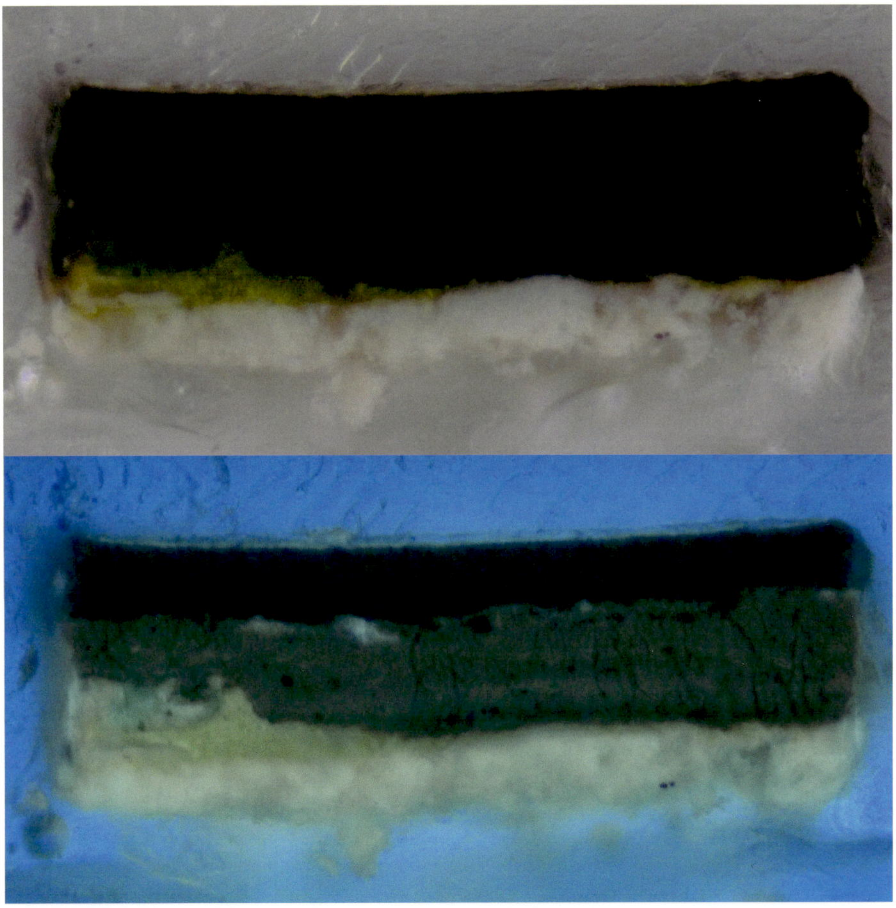

First-Class Marksman and *The Camp* also reveals lower paint films with ridges of drying wrinkles. This black paint wrinkling is similar to that observed during drying of the gloss black Ripolin from the can and is likely an inherent defect in the formulation of the paint; it perhaps explains Nolan's repainting them with a nonwrinkling alkyd.

The difficulty of the slow drying and formation of wrinkles in black enamel paints was overcome in the 1930s by the development of alkyd resins. Alkyds dried more quickly and could be formulated without the need for surface driers that formed a skin under which the paint continued to shrink and wrinkle. Alkyd paints would also have been easier to control, enabling Nolan to achieve the sharp edge of the armor while retaining the desirable high gloss finish for the Kelly figure. Indeed, the use of an alkyd paint on the Kelly black armor belies traditional assertions that Nolan used Ripolin exclusively on these paintings.

By July 1947 the Kelly theme had, for the moment, worked itself to an end, and Nolan was looking for new subjects and inspiration for his paintings. He decided to leave Heide and travel north to Queensland, a turn that took him permanently away from the Heide influence and on to a new and unexpected artistic pathway and relationships.

FIGURE 49
Plumber's bag of Ripolin paint in adapted cans from Nolan's Wahroonga studio. Artists' Materials Archive, Art Gallery of NSW.

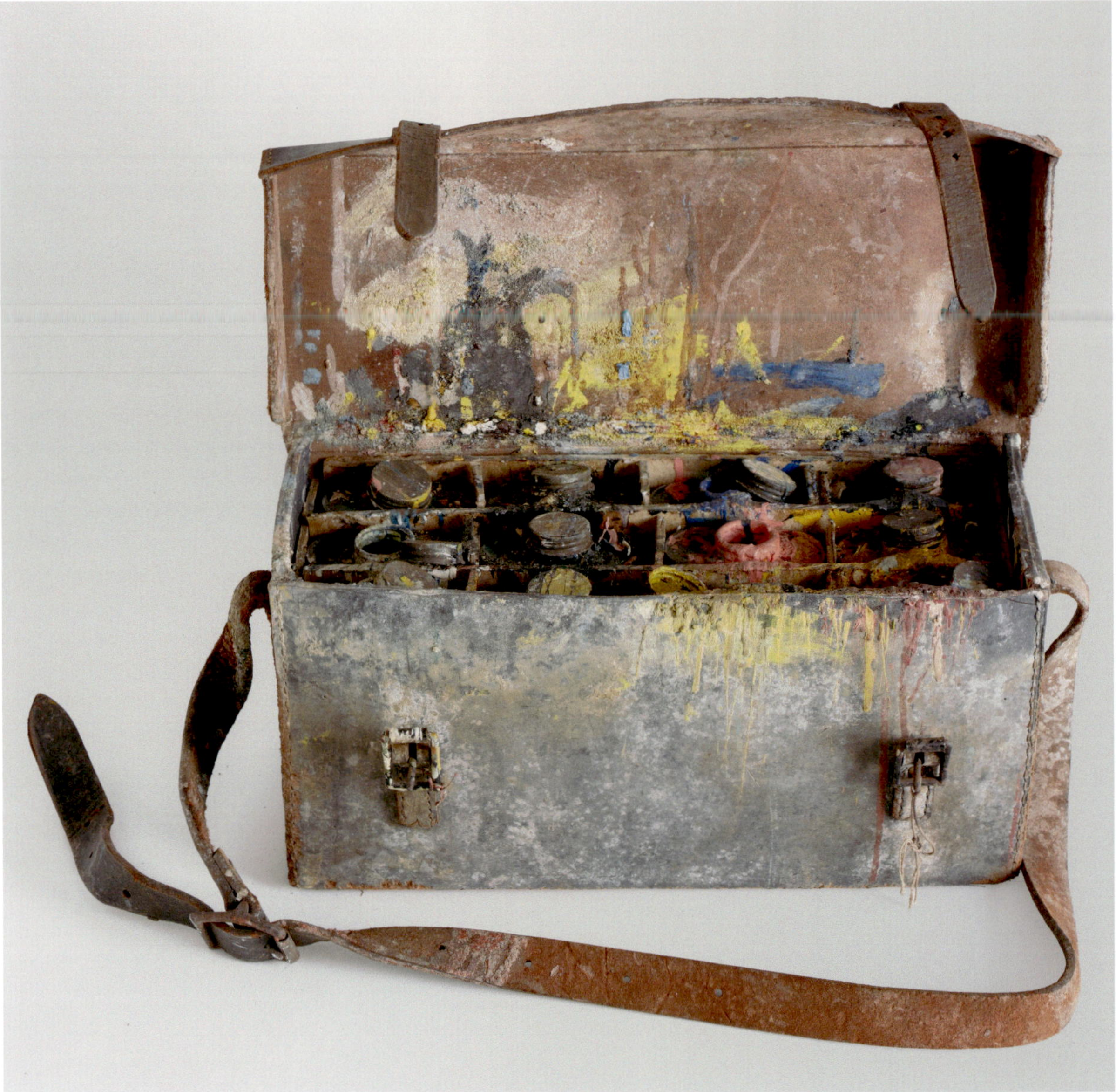

4 Queensland Journeys: 1947

Paula Dredge, Anne Carter, and Gillian Osmond

Nolan finished painting his Ned Kelly series at the beginning of July 1947, after which he and the Reeds decided that he should take a holiday in Queensland. Nolan had met the poet Barrett Reid at Heide during Christmas 1946, and he now accepted Reid's invitation to stay with him at his father's house in Brisbane (Reid 1967, 447). No major pronouncements preceded this decision; however, the Reeds had just begun full-time care of Sweeney Tucker, the three-year-old son of Albert Tucker and Joy Hester, and it seems there was a general desire for change (Harding and Morgan 2015, 196). Nolan left Melbourne on July 8, 1947.

One of Nolan's first letters to Sunday Reed from Brisbane, dated July 14, describes his pleasure in flying in an airplane and seeing the country from the air, and his renewed interest in the depiction of the horizon and distance:

> Looking straight down from the plane gave me that questioning I felt on first seeing the flat Wimmera from the back of a motor truck. In that case the way to paint it was to put the flat landscape vertically up the canvas. The position is exactly reversed from the plane. How to solve it I do not know as yet but modern painting has probably been too much in the hands of old-fashioned men.[1]

Three days later, Nolan wrote to Sunday of a mysterious box that he had brought along: "The chap at the plane door told me to put the black box under my seat & not up on the rack as parcels had been falling on passenger's heads. Little did he know when he smiled just what would have happened had the innocent black box started floating around the aircraft." The black box was probably Nolan's traveling paint satchel, now in the Artists' Materials Archive at the Art Gallery of NSW, gifted, along with the paints Nolan left in his Wahroonga studio when he left for London in 1953, by Jinx Nolan in 2006 (fig. 49). It is a rigid leather plumber's bag filled with a dozen screw-topped containers of decanted Ripolin paint. The outside of the containers are painted khaki and had a metal belt attachment that has been removed and filed smooth. They are stamped with a manufacturing date of 1943 and the name of the maker, United Can Company (UCC), which supplied the Australian Army during the war. These adapted army containers provided an inventive and resourceful strategy for traveling with the Ripolin paints. Unlike typical artist's paints, Ripolin was too liquid and fast-drying to be stored in tubes, and the cans that Ripolin came in were no doubt too heavy to carry and awkward to work with outdoors. The easily opened screw tops would have allowed ready access to the full range of paint colors, as well as quick sealing to retain

FIGURE 50
Sidney Nolan, *Female Nude*, 1947. Monotype, 25.2 × 31.5 cm (10 × 12½ in.). Heide Museum of Modern Art. Bequest of Barrett Reid 2000.

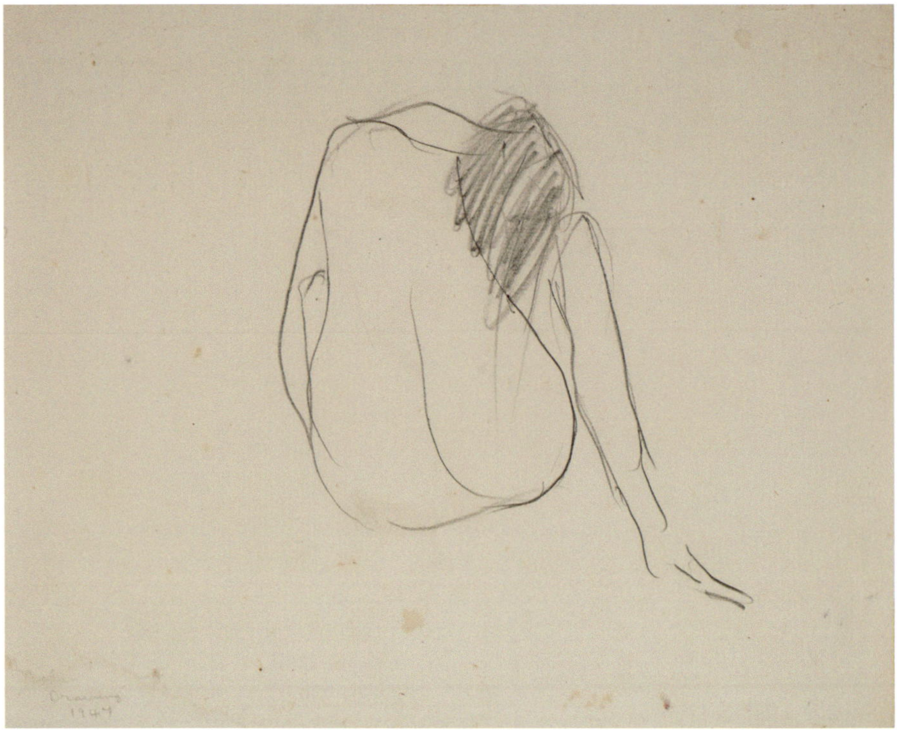

the volatile solvents. Nonetheless, it still seems to have been a weighty burden; as Nolan remarked to Albert Tucker from Brisbane on July 16, "My 30 lbs of Ripolin looks me in the eye every morning, soon have to start making it lighter" (McCaughey 2006).[2]

Analysis of the paint in the travel satchel shows good binder and pigment correspondence with the cans of Ripolin paint from Nolan's Sydney studio (see appendix II, table 3).[3] The satchel contains twelve containers of paint: four white, two Prussian blue (one deep and one pale blue), one phthalocyanine blue (likely Ultramarine 13, which did not in fact contain ultramarine pigment), one green, two yellows (Lemon Yellow and the other corresponding to Canary Yellow), one black, and one pale red. All are oil-based. Just one color, the pale red, has pigments not found in the Ripolin cans from the studio.

During his first weeks in Brisbane, according to Barrett Reid, Nolan did little but eat pawpaws and climb mountains. Still, he proved to be a challenge for his hosts. "We had to keep up the tradition of northern hospitality," Reid noted. "After all, we fed the American army. But that was nothing, compared to him."[4]

Nolan may have been a demanding guest, but he was also a breath of life in Brisbane. He was introduced to artists at the Barjai-Miya studios and became reacquainted with the poet Charles Osborne, whom he had met previously at Heide and whose family ran the Ballad Bookstore, where Nolan attended life-drawing classes and drew works such as *Female Nude* (1947; fig. 50).

Yet Reid felt that Nolan was distracted, that Brisbane was "but the absence of Ned Kelly and all the things around his life in Melbourne." And he was not painting much. "Apart from the lovely birds on glass which were done before the first Frazer [*sic*] Is. trip, Nolan did a few little canvases, all in the last week, and I think it may have surprised him a little how stiff his arm had become."[5]

Although the plan for Nolan to find new subject matter in Queensland was slow to evolve, he did travel extensively. First, he retraced the Reeds' 1946 holiday itinerary, traveling to Green Island on the Great Barrier Reef and to the far-northern Queensland towns of Cooktown and Laura (Underhill 2015, 184). Returning to southern Queensland, he then joined a group from the Queensland Geographical Society on a trip by horseback to Carnarvon Gorge to see the spectacular rock art on Bidjara and Garingbal land.

Nolan took photographs during these travels, despite the fact that "films are hard to get here."[6] The photographic prints he sent to the Reeds in Melbourne include several of the Carnarvon rock art,[7] and he later used many of the Queensland photographs as studies and inspiration for paintings.[8] At the time, however, Nolan was focused on another Queensland subject, Fraser Island, which culminated in a series of twelve paintings dated October 1947 and exhibited in Brisbane in 1948.

Painting on Fraser Island

Fraser Island, north of Brisbane, is the largest sand island in the world. Its national park has recently been returned to its Aboriginal name, K'gari. It has an incredibly lush interior of tall forest and clear-water white-sand lakes and creeks, and an inner coastline with stands of tall swamp banksias. On the ocean side, the coast is wild and windswept, with large sand dunes called sand blows, pushed inland across the hills and trees behind the beach.

Nolan had become interested in the story of the shipwreck of the *Stirling Castle* off the northern tip of Fraser Island in 1836 and its sole survivor, Mrs. Eliza Fraser, the captain's wife. Accounts of Mrs. Fraser's experience on the island after the wreck became legendary through multiple retellings. The accounts available to Nolan in the 1940s were enthralled by Eliza's apparent capture and humiliation at the hands of the local Butchulla people and her "rescue" by an escaped convict, David Bracewell. She was said to have betrayed Bracewell by reneging on her promise to petition for his freedom on her return to Brisbane (Seear 1998). Newer histories, however, reveal that the local people, rather than humiliating her, in fact soothed Mrs. Fraser's sunburn and protected her during her weeks on the island, saving her from certain death (Behrendt 2016).

Postwar Fraser Island was host to a timber industry based in several camps in the densely wooded interior of the island. The pristine coastal landscape, the culture of the timber workers, and especially the story of shipwrecked Eliza Fraser captivated Nolan, perhaps in part because of his own fugitive status with the Australian Army. Fraser Island was a place "where you go to hide something—there were a lot of dodgy characters" (Rafferty 2016). Nolan had it in mind to "follow the story of the convict & Mrs Frazer [*sic*] on their journey down from Frazer [*sic*] Is to Brisbane" (Nolan to Sunday Reed, July 15, 1947), and although access to the island was difficult in the postwar years, he was determined to make the journey. In early August he described sailing past Fraser Island, taking note of the spectacular sand blows along the coastline, but he was unable to immediately find a way ashore. Later that month he successfully navigated a short visit. However, it was a longer stay, from September 29 to October 24, that at last gave him the opportunity to paint.[9]

In a letter to John Reed of February 3, 1948, Nolan confided that the paintings done on Fraser Island were a struggle and that none really represented what he wanted to say about the place.[10] This struggle may have been tied to his feelings for Sunday Reed, for it was between his visits to Fraser Island in August and his return in September that Nolan resolved not to resume their romantic relationship. This break was not easily arrived at, however, and was very much tied into his Fraser Island experience; as he wrote to John Reed on October 23, 1947, it was "true that in many ways I feared the impact of the

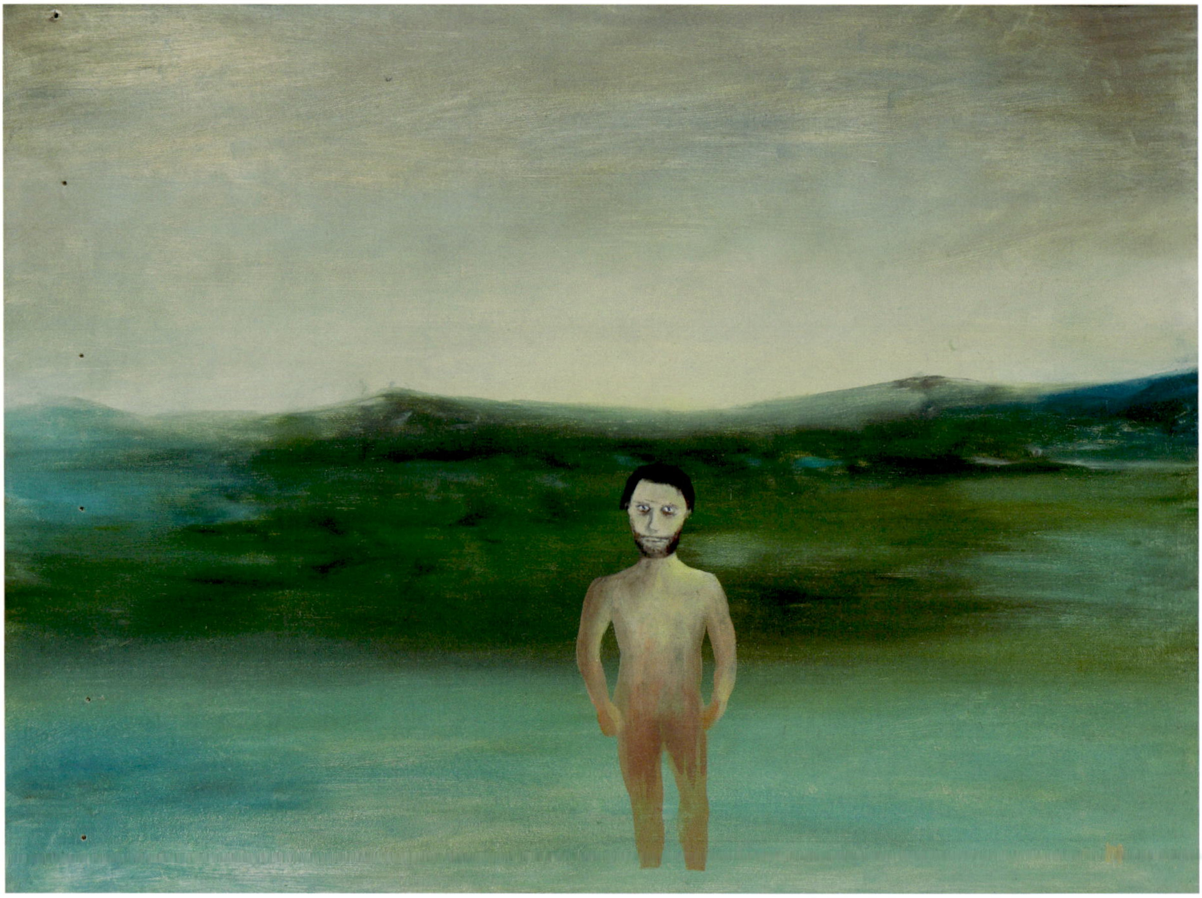

FIGURE 51
Sidney Nolan, *Island*, 1947. Ripolin oil enamel on Masonite hardboard, 77.0 × 105.0 cm (30¼ × 41⅜ in.). Art Gallery of NSW. Purchased with funds provided by the Art Gallery Society of New South Wales 2001.

island, sensing that if I stayed on in the silence my emotional structure would crack. And so it did, & so it had to."

Nolan described his first visit to Fraser Island in detail in a letter to John Reed dated August 28, 1947.[11] He had traveled across to the island in a punt used for transporting cut timber and was then driven by the timber workers to various parts of the island in an open truck.

> Saturday morning they filled up the big four-wheel drive truck and set off 40 miles along the ocean coast northwards to Indian Heads which is about 17 miles from the top tip of the island. A two day trip and families and tents and food all in the back of the open truck. It was a lovely break, apparently it is only done once in a blue moon. The truck drove along the hard sand at low tide and a glorious pure white ocean beach stretching back and front as far as you could see. Trickling down from the cliffs were cracks which carved out a channel through the sand between each tide and the fresh water and salt met at an exact point. Some of the sandstone cliffs had eroded into carved towers and pinnacles with all the colors under the sun mingling into each other with each layer of sandstone. Pinks and yellow ochres & crimsons. Hundreds of brumbies [feral horses] moving in the hillsand occasionally groups of them galloping along the beach. The most spectacular part of the coast however are the huge sand blows which have blown up from the beach during the years and cover whole hills. These were the huge patches of sand I had noticed from

> the boat. At Indian Head where we stopped the night, is the largest blow of all, about five miles long and covering the hills back for a mile in some places. Down between the sandhills there was nothing to distinguish it from what the Sahara must be. You can imagine the feeling of the place when this takes place right next to the ocean.

On September 29, Nolan returned to spend more time on Fraser Island. He was by then planning an exhibition at the Moreton Galleries in Brisbane for the following year. On December 5 he wrote to John Reed, "I did paint hard in the latter period on Frazer Is & there are a dozen fairly big paintings that will be more than enough I think for the exhibition in Febuary [*sic*] with Cooper. I took photographs to send down but so far have not had the prints from Bris." For this second trip, Nolan stayed at Central Camp with Norm Crombie, the forestry overseer; his wife, Valda; and their nine-year-old daughter, Anita. He set up a makeshift studio on the veranda and got to work painting (Rafferty 2016).

Nolan described Norm as "an interesting bloke, young chap about thirty, half bearded..., wild eyes and moves about in bare feet like an animal.... In many ways he is next door to what the bushrangers must have been like" (Nolan to John Reed, August 28, 1947). The painting *Island* (1947; fig. 51) depicts Norm standing in one of the clear-water sandy-bottomed lakes. Built up in thin atmospheric washes of Ripolin in blue, pink, and green over a white ground layer, the painting merges the water, land, and sky; in mood and technique, it is a striking departure from the high-keyed color and densely worked paint of the Wimmera and Kelly country landscapes. The subject's feet disappear into the white sand beneath the water, and his intense blue eyes mirror the color of his surroundings.

Nolan's Salvaged Masonite on Fraser Island

In postwar Brisbane, painting materials were difficult to obtain; even paper was in short supply.[12] On July 24, 1947, following his arrival in Brisbane, Nolan had written to John Reed asking him to send supports: "The prepared boards & canvases 25" × 30". Canvas for preference...had better start getting used to canvas, Masonite looks hopeless here for months." These materials had arrived by August 9, but they were not used for the Fraser Island paintings.

Norm Crombie cheekily promised to cut Nolan some tree bark to paint on,[13] but instead Nolan used his favorite painting support, Masonite. A review of the Fraser Island exhibition was critical of Nolan's "deliberate maltreatment of so much useful and hard-to-come-by building material" (Webb 1948), as postwar Queensland was still suffering considerable material shortage, but this criticism, Nolan complained, was unfair. He had, he said, retrieved much of the Masonite from a water-logged, termite-infested heap on the island (Nolan to John Reed, February 26, 1948).[14] In support of this assertion, Anita Crombie recalled Nolan disappearing from their house at Central Camp, only to return later with boards on which to paint (Rafferty 2016).

It has been possible to examine the Masonite boards used in six of the twelve works from the Moreton Galleries exhibition to test Nolan's assertion of their source.[15] The works are *Island* (Art Gallery of NSW; fig. 51), *Lake Wabby* (Heide MoMA; fig. 52), *Platypus Bay, Fraser Island* (*Platypus Bay,* QAGOMA), *Mrs. Fraser* (*Woongoolbver Creek,* QAGOMA; fig. 53), *Fraser Island* (private collection; fig. 54) and *Swamp* (private collection; fig. 55).[16] A seventh painting from the Moreton Galleries exhibition,

FIGURE 52
Sidney Nolan, *Lake Wabby*, October, 1947. Ripolin oil enamel on Masonite hardboard, 77.3 × 106.3 cm (30½ × 41⅞ in.). Heide Museum of Modern Art.

FIGURE 53
Sidney Nolan, *Mrs. Fraser* (originally titled *Woongoolbver Creek*), 1947. Ripolin oil enamel on Masonite hardboard, 65.8 × 106.6 cm (26 × 42 in.). Queensland Art Gallery. Purchased 1995 with a special allocation from the Queensland Government, celebrating the Queensland Art Gallery's Centenary, 1895–1995.

FIGURE 54 (opposite top)
Sidney Nolan, *Fraser Island*, 1947. Ripolin oil enamel on Masonite, 76.0 × 104.0 cm (30 × 41 in.). Private collection.

FIGURE 55 (opposite bottom)
Sidney Nolan, *Swamp*, 1947. Ripolin oil enamel on Masonite, 74.5 × 104.0 cm (29⅜ × 41 in.). Private collection.

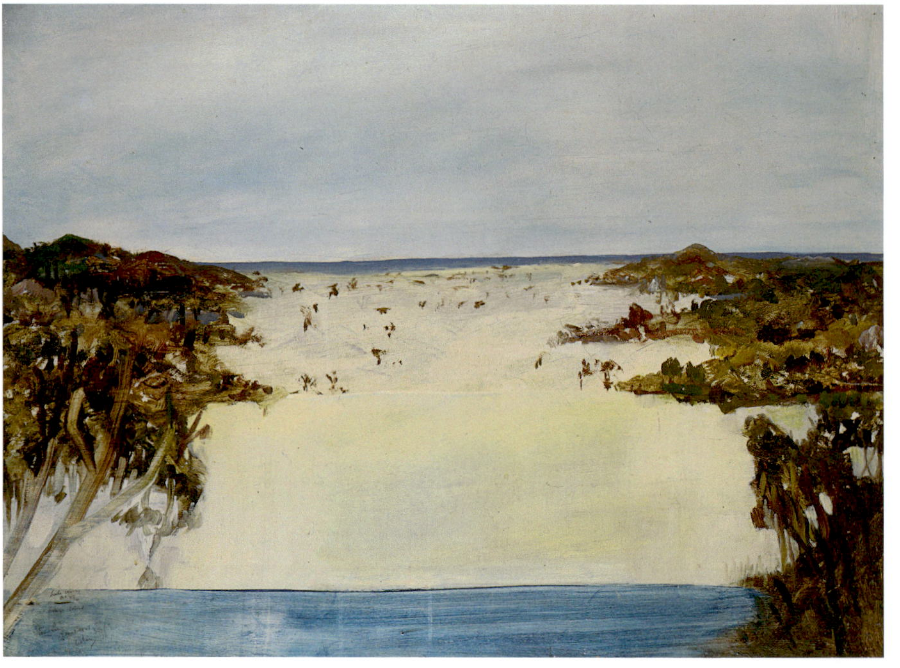

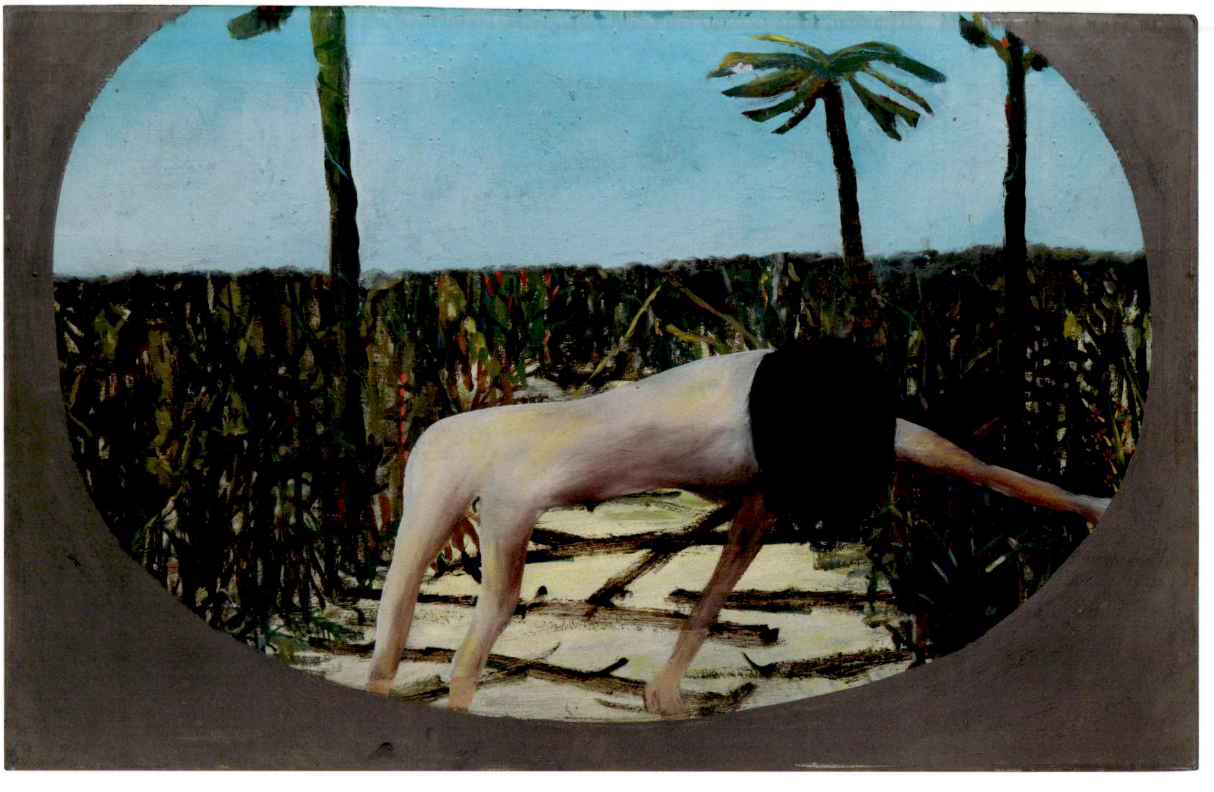

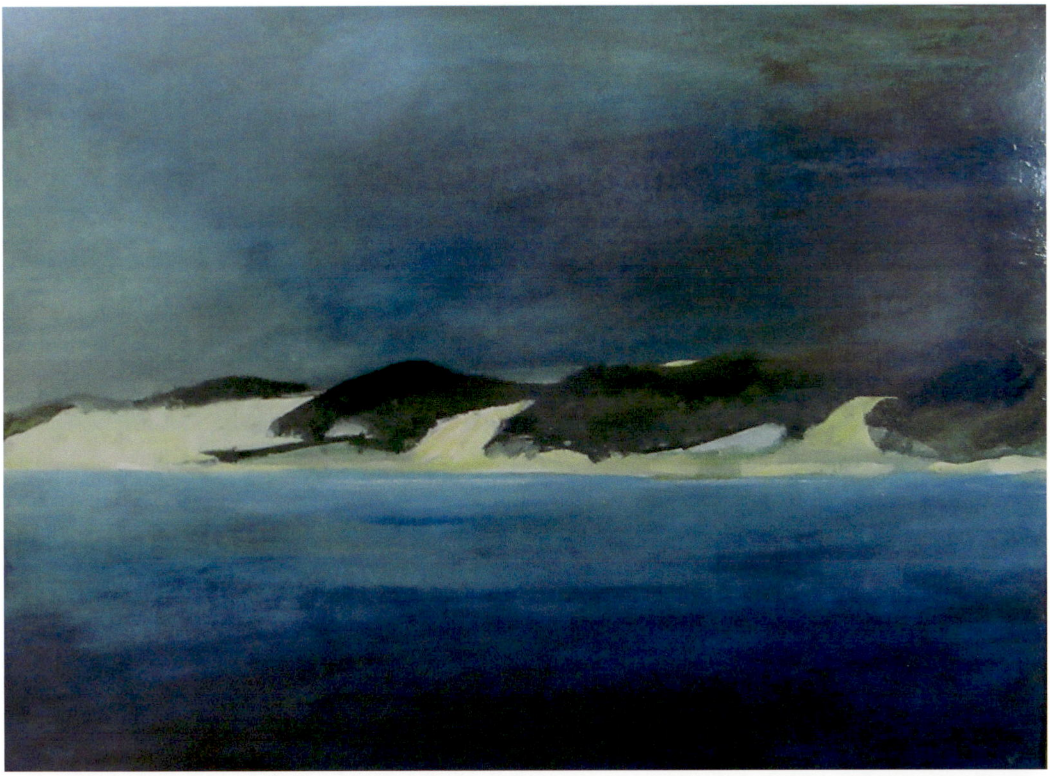

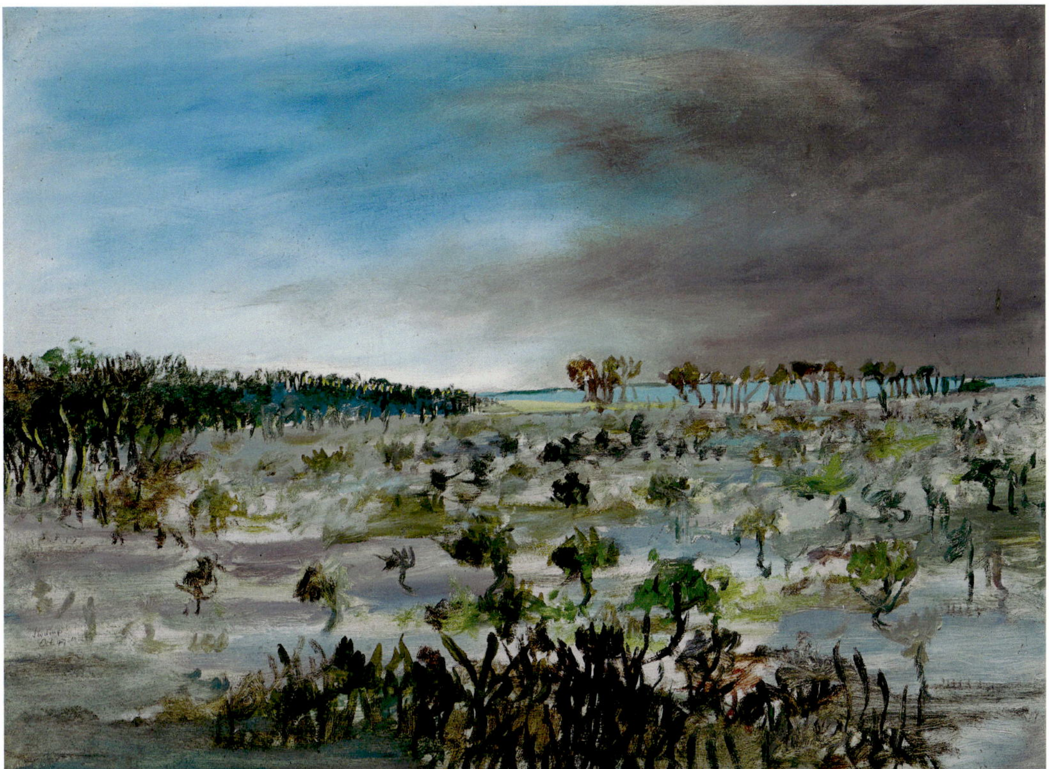

Indian Head, appears to have been destroyed by Nolan, but a remnant off-cut comprising approximately the lower third of the original painting was made available for this study (private collection; fig. 56). The four paintings in public collections and the remnant from *Indian Head* have been analytically tested (see appendix I, table 2). Inspection of the reverse of all seven examined hardboard supports confirms Nolan's use of secondhand Masonite. *Island, Lake Wabby,* and *Indian Head* show preexisting nail holes, which Nolan took no trouble to fill or disguise before painting. The back of *Island* is coated in a powdery white paint that is missing from an area in line with the nail holes, suggesting that it was painted while nailed to a support, probably the framework of a building. White paint is also present on the back of *Mrs. Fraser, Platypus Bay, Swamp,* and *Indian Head;* these white paints all have the same pigment composition, consistent with a popular brand of water-based interior paint called Kalsomine™.[17] *Fraser Island* and *Lake Wabby* have no white paint on the reverse; however, *Lake Wabby* does have a "tempered presdwood" stamp, indicating its production as a Masonite wet-area product. The paintings are all executed on the smooth side of the Masonite, and the backs show signs of water damage and staining. Five paintings are similarly sized with approximate dimensions of 77 × 106 cm (30¼ × 41¾ in.).[18] They are smaller than Nolan's usual (Kelly) Masonite boards, which were prepared from a standard 91 × 366 cm (48 × 144 in.) board cut to produce four boards each 91 × 120 cm (48 × 36 in.), suggesting that the edges of the Fraser Island panels were all trimmed either to fit architectural wall sizes or to remove edge damage after they had been salvaged by Nolan.

Further evidence of the recycled nature of the Fraser Island Masonite boards is the presence of a dark green paint layer beneath Nolan's paint on all of the paintings examined. The composition of the green layer was found to be consistent in all but one of the five paintings analyzed with FTIR spectroscopy.[19] The binder is probably an oil-based paint. Pigments identified are Prussian blue and chrome yellow, together with barites and zinc white. The dark green color made from this mixture of Prussian blue and chrome yellow is commonly called Brunswick green and was a typical house paint and army color of the period.

On all of the examined Fraser Island paintings, the green paint layer contains fibrous particles that give the finished surface a distinctive irregular texture. On *Mrs. Fraser,* moreover, this layer is roughly applied, with thick paint drips running down the face of the panel. Visible in the upper right of the X-radiograph of the painting (fig. 57), the drips run in the opposite direction to Nolan's painting, suggesting that he rotated the Masonite from its original orientation as a building board painted green. In a micro paint sample prepared as a cross section (fig. 58), dirt particles can be seen on top of this existing green layer, indicating elapsed time between this layer and subsequent paint layers.

FIGURE 56
Remnant (lower section) of Sidney Nolan, *Indian Head,* 1947. Ripolin oil enamel on Masonite hardboard, 27.6 × 106.4 cm (10⅞ × 41⅞ in.). Private collection, courtesy of Sotheby's Australia.

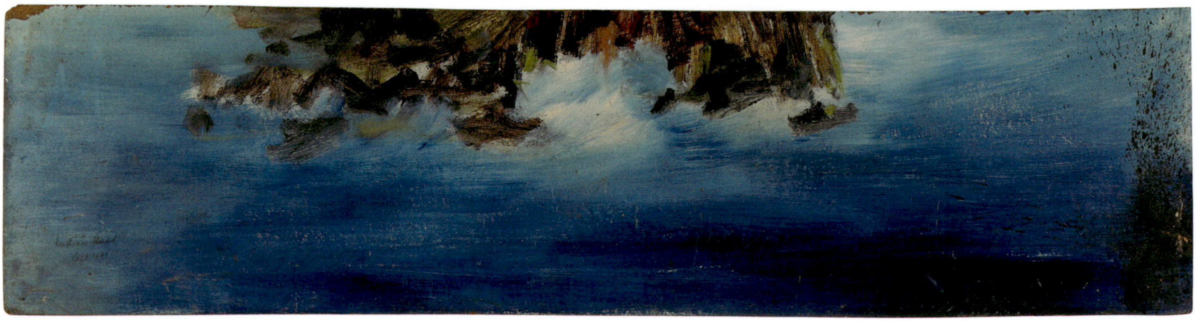

FIGURE 57
X-radiograph of *Mrs. Fraser* (see fig. 53) showing the continuation of foliage beyond the current horizon line and up to the present top edge. Mrs. Fraser's feet are also visible beneath the oval border. Vertical drips across the right side of the painting relate to a layer of green paint that predates Nolan's composition.

FIGURE 58
Micro paint sample taken from edge of *Mrs. Fraser* (see fig. 53) in incident light (top) and ultraviolet light (bottom) showing build-up of paint layers: Masonite (A), green paint (B), white priming layers (C and E) with zinc white-based interlayer (D), green and blue layers (F and G), and gray oval border (H).

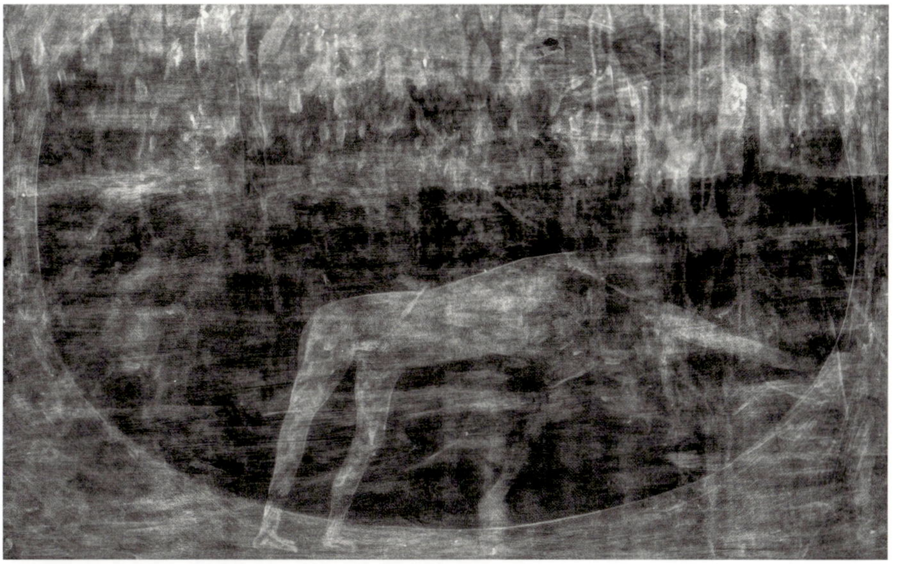

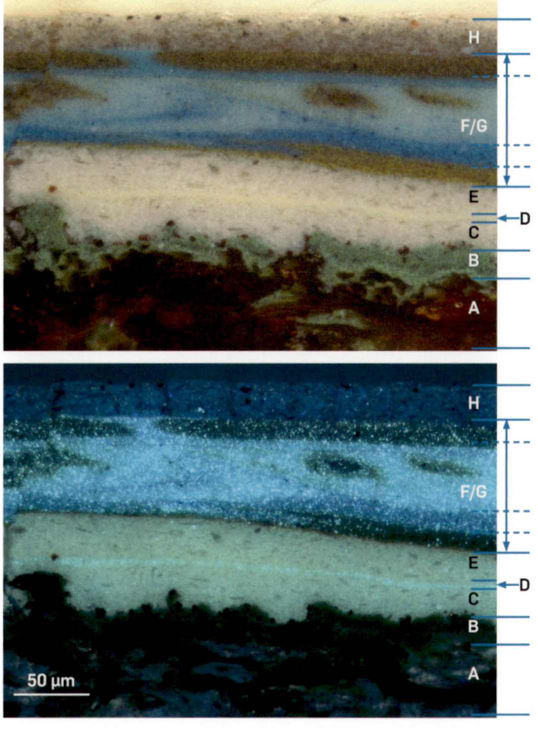

The Masonite that Nolan salvaged from "white ant ridden heaps" on Fraser Island has a likely source. During the Second World War, over a thousand soldiers from the Fraser Island Commando School, or Z-Special Unit, occupied a camp around Wanggoolba Creek on the west coast. An inventory made prior to the dismantling and dispersal of the camp in 1945 details buildings with Masonite walls, including the butcher and hospital.[21] Almost all the Commando School buildings were painted green. It is thus quite feasible that, just two years after the camp was torn down, Nolan was able to retrieve remnants of the buildings for painting supports.

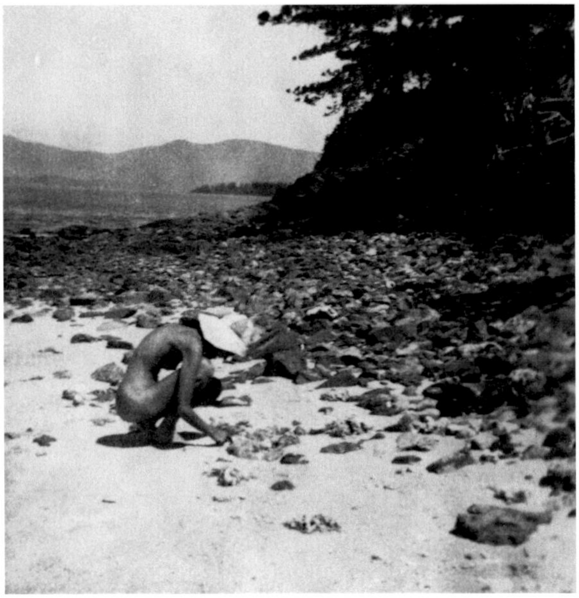

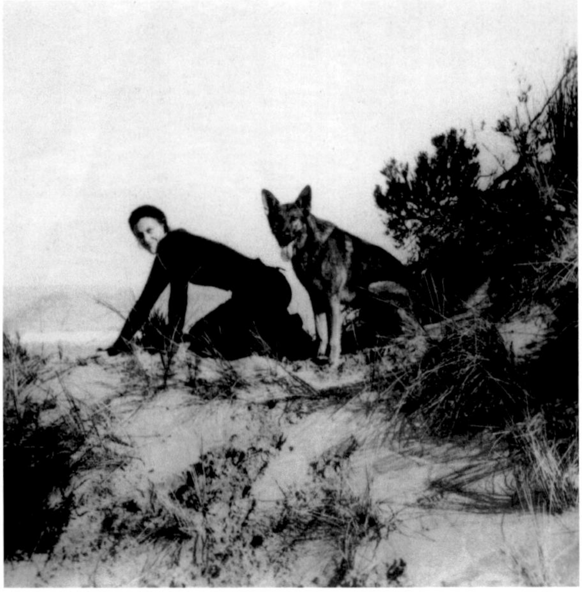

FIGURE 59
Photographer unknown, Sunday Reed at Green Island, Queensland, 1946. Heide Museum of Modern Art Archive.

FIGURE 60
Photographer unknown, Sunday Reed and Karel, c. 1932. Heide Museum of Modern Art Archive.

Although in all seven of the examined paintings the green pre-painting was covered with a white priming layer, the general tonality of the Fraser Island paintings is dominated by variations in greens with a misty representation of light. This new color and light in Nolan's repertoire was no doubt inspired by the lush vegetation of the island itself, but the green boards from the bush rubbish heap must have also influenced his choice. They were a remarkably serendipitous find.

Mrs. Fraser, 1947

The most iconic painting from the Fraser Island series was exhibited in 1948 as *Woongoolbver Creek* and later retitled *Mrs. Fraser* by Nolan (see fig. 53). It is an animal-like depiction of Mrs. Fraser on all fours in Wanggoolba Creek (as it is now spelled), her face hidden by her hair—much like the figure Nolan drew in Brisbane at the Ballad Bookstore (see fig. 50). A photograph of Sunday Reed nude on the beach during her Queensland holiday the year before (fig. 59), and an earlier photo of her and her dog Karel from 1932 (fig. 60), also exhibit striking similarities with the Mrs. Fraser figure.[22] Nolan would certainly have seen the Queensland holiday photo, after John and Sunday's return to Melbourne and before his own trip north the following year, and very likely the photo of Sunday and Karel as well, during his long stay at Heide. Unlike any of the other works from the Fraser Island group, and in contrast to Nolan's usual practice of working just once on a painting, *Mrs. Fraser* went through at least six revisions at Nolan's hands between 1947 and 1957. Some of these may have been in response to the critical reception it received in 1948; other modifications may have been motivated by his changing relationship with Sunday Reed. Unpicking the sequence of the physical changes may help us to understand how this painting in particular retained such a long fascination for Nolan.

While no photograph of *Woongoolbver Creek* from the 1948 Moreton Galleries exhibition has been found, Nolan photographed the painting in late 1947 along with *Foreshore* and sent those photos together with photos of his travels in Queensland to the Reeds. *Woongoolbver Creek* is photographed leaning against a wall, with a large expanse

FIGURE 61
Sidney Nolan, *Woongoolbver Creek* (now titled *Mrs. Fraser;* see fig. 53) leaning against wall, 1947. Silver gelatin photograph, 6.5 × 9.0 cm (2⅝ × 3⅝ in.). State Library of Victoria, Melbourne.

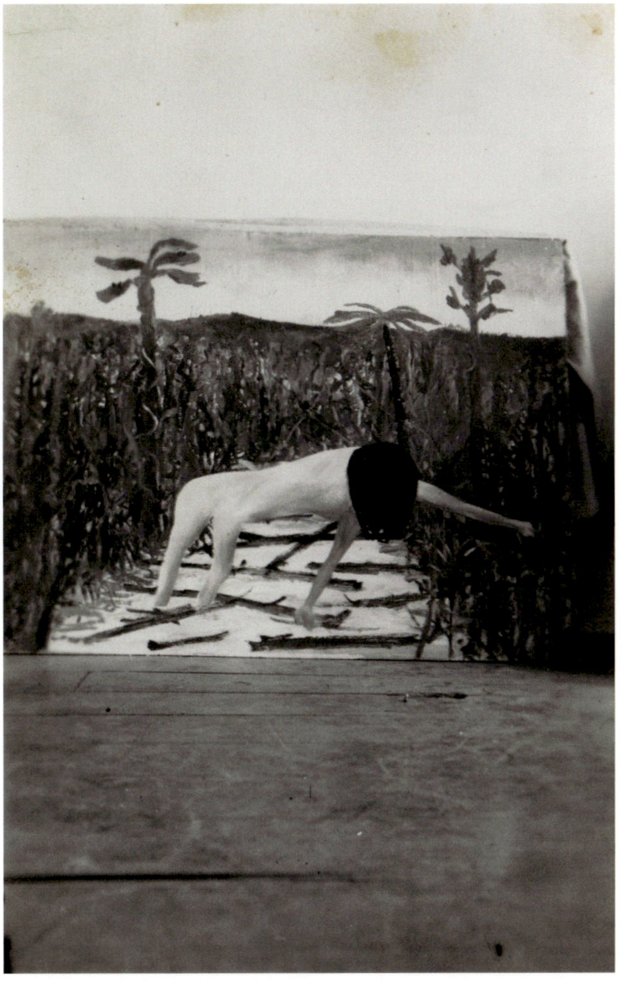

of floorboards in the foreground (fig. 61).[23] This is likely how *Mrs. Fraser* appeared when it was exhibited in 1948, but it has since undergone substantial changes, with Nolan lowering the horizon line, adding a gray ocular border, and trimming the height of the Masonite support. The sequence and dating of the changes have been subject to ongoing speculation (Rainey 2012; Underhill 2015), although they were certainly complete by the time of Nolan's 1957 Whitechapel Gallery exhibition in London, when the painting was featured in a number of photographs of Nolan in his London studio surrounded by his materials and other paintings.

Tiny samples were taken from the edges of the painting. These were cut in cross section to reveal the layering sequence of the paint. A total of seven layers are visible in cross sections from *Mrs. Fraser,* revealing not only that Nolan painted priming layers over previous layers to "start again," but also that he used a mixture of paint types. The micro paint sample in cross section (see fig. 58) provides important clues to Nolan's manipulations of *Mrs. Fraser.* First, directly on top of the preexisting green paint layer (B in fig. 58), there is a white priming layer (C), and over this a thin wash of colored paint, seen in samples from different areas as blue, gray, or, as in figure 58, cream (layer D). Under ultraviolet illumination in paint cross section, this thin layer is distinctive in the fluorescence of its zinc white particles, in contrast to the priming layer. The fluorescence is the first clue that links this paint layer (D) to subsequent image layers (F and G) and suggests Nolan's use of Ripolin, which also contains fluorescent zinc white particles.

FIGURE 62
Photomicrograph of *Mrs. Fraser* (see fig. 53): wet paint of the shoulder meets the wet paint of the hair.

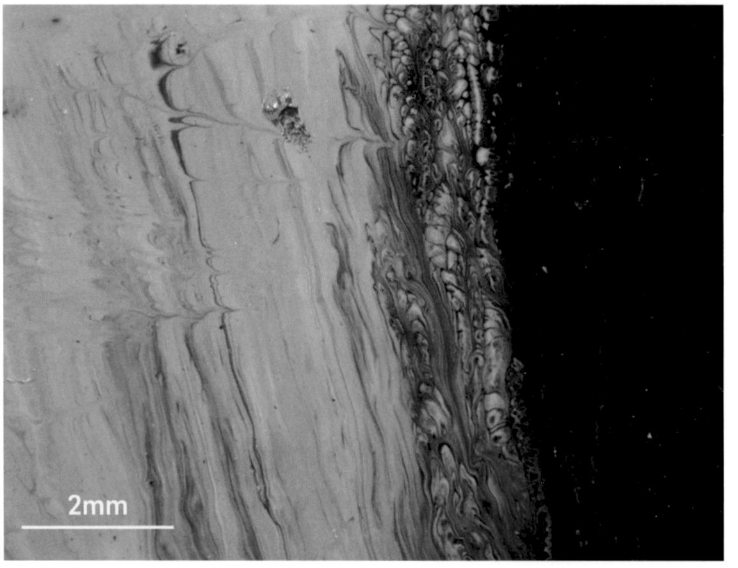

This thin layer of Ripolin was then sandwiched by a second nonfluorescent priming (layer E), visually identical to the first white priming. Perhaps Nolan rejected the previous image and applied a second white priming layer to allow for a fresh start. The covered paint is very thin, likely an initial washlike laying-in of background, consistent with his other paintings from Fraser Island. The composition of the two priming layers on *Mrs. Fraser* is also identical with that on the other Fraser Island paintings that have been analyzed (*Platypus Bay, Island, Lake Wabby,* and *Indian Head*), suggesting that they were prepared contemporaneously.

Nolan's second campaign of painting on *Mrs. Fraser* began as a landscape, with the creek and vegetation having a distinct red tonality. At this stage, as both X-radiography (see fig. 57) and infrared reflectography (IRR) reveal, the vegetation continued almost up to the top edge, and there may have been only a narrow band of sky. Converging lines that relate to the perspective of the creek are present in the X-ray image and remain visible in the painting below the horizon, their vanishing point at the original center of the Masonite support. Photographic details of the paint surface show the extraordinary qualities of the Ripolin when worked quickly, with color edges flowing together to form marbling effects. That Nolan was able to work with this uncontrollable element of the paint and embrace its accidental effects shows remarkable courage and versatility.

IRR and surface examination indicate that the figure of Mrs. Fraser was not present at this early stage, but was painted later over the landscape. The addition of her body appears to be a third working, completed before the 1947 photograph of the work was taken. The figure has been painted quickly; the hair merges into the wet paint of the shoulder and arm, producing mingled paint effects and demonstrating the simultaneous working of the body and the hair (fig. 62). There is no evidence that Nolan ever painted a face, as the landscape is visible between the brushstrokes of the black hair. Mrs. Fraser's hair is an oil-based paint with an assumed carbon-based black together with small amounts of lead chromate and Prussian blue, consistent with Ripolin black gloss paint. Her black hair forms a mask, a visual device similar to Kelly's helmet and an effective strategy for depersonalizing the figure, suggesting isolation and internal withdrawal. Another change introduced by Nolan during this third working includes lowering the sky. The green foliage of the underlying paint remains visible where blue sky has been painted around the edges of palms.

Nolan made additional changes to the painting sometime after its exhibition in 1948 but prior to its reshowing in the Whitechapel Gallery in London in 1957. In a fourth working, the horizon was lowered a second time, and feet were added to Mrs. Fraser where previously they were absent (see fig. 61). Sometime after these changes had dried, a fifth session added a gray, oval-shaped border (see layer H in fig. 58), framing the work in an elliptical optic lens that produced a disturbing sense of observation and objectification. This border hid Mrs. Fraser's feet from view, but they remain clearly visible in the X-radiograph (see fig. 57). In contrast to the gloss of the rest of the painting, the gray border paint is matte and easily marked, showing ghosting of fingerprints and many scratches. FTIR analysis indicates it is an alkyd medium containing kaolin, silica, talc, and barites. The black pigment is presumed to be carbon-based. This alkyd medium is thus distinguishable from the oil-based Ripolin used in the earlier campaigns of painting. Because no mention is made of the gray border in the reviews of the 1948 exhibition, it is presumed that its addition postdates that show, though whether done in Sydney or London is not known. The final revision, a sixth working stage, involved cutting the top eleven centimeters (4½ inches) from the Masonite panel to foreshorten the point of perspective. A thick pencil line and fractured paint at the top edge indicate that a line was drawn on the dried paint and the board was cut after all the paint layers, including the gray border, were dry.

Paint on the five Fraser Island paintings analyzed demonstrate zinc oxide–like fluorescence and the consistent presence of zinc carboxylates in FTIR spectra (see sidebar on p. 71: Ripolin and Zinc Soaps). These features, combined with pigment identification consistent with the pigments found in the Ripolin from Nolan's studio and his traveling bag, indicate (with the exception of his white priming layers and the gray border) paints characteristic of Ripolin. In four paintings—*Island, Platypus Bay, Lake Wabby,* and *Indian Head*—one white priming layer was applied over the preexisting green paint and image layers were worked up in a single painting session, in contrast to the more extensively modified and reworked *Mrs. Fraser.* It would seem, therefore, that *Mrs. Fraser* alone of the Fraser Island paintings was subject to Nolan's reworking and reconfiguration.

Sydney

By the end of 1947 Nolan was ready to leave Brisbane. He vacillated between returning to Melbourne or stopping in Sydney, but eventually decided on Sydney. After spending Christmas with close friends, the artists Joy Hester and her husband Gray Smith, Nolan met up with John Reed's sister Cynthia and her daughter, Jinx, who were living in Wahroonga, a suburb on the northern side of Sydney Harbour. He soon moved in with Cynthia and Jinx, and in March 1948 he and Cynthia married. This new relationship caused a rift with John and Sunday Reed that was not reconciled in their lifetimes and indeed became increasingly acrimonious over the following decades.

The twelve Fraser Island paintings were exhibited at the Moreton Galleries, Brisbane, in February 1948. The gallery director, John Cooper, was supportive of Nolan and rearranged his schedule to fit in the late-notice exhibition, despite it being a commercial risk (Skinner 1984, 38). The show was unlike anything seen in Brisbane before. The paintings were presented on easels rather than hung on the walls (fig. 63) and were framed in light-colored two-inch Queensland pine frames. Thenceforth, Nolan continued to use local timbers to frame his paintings.

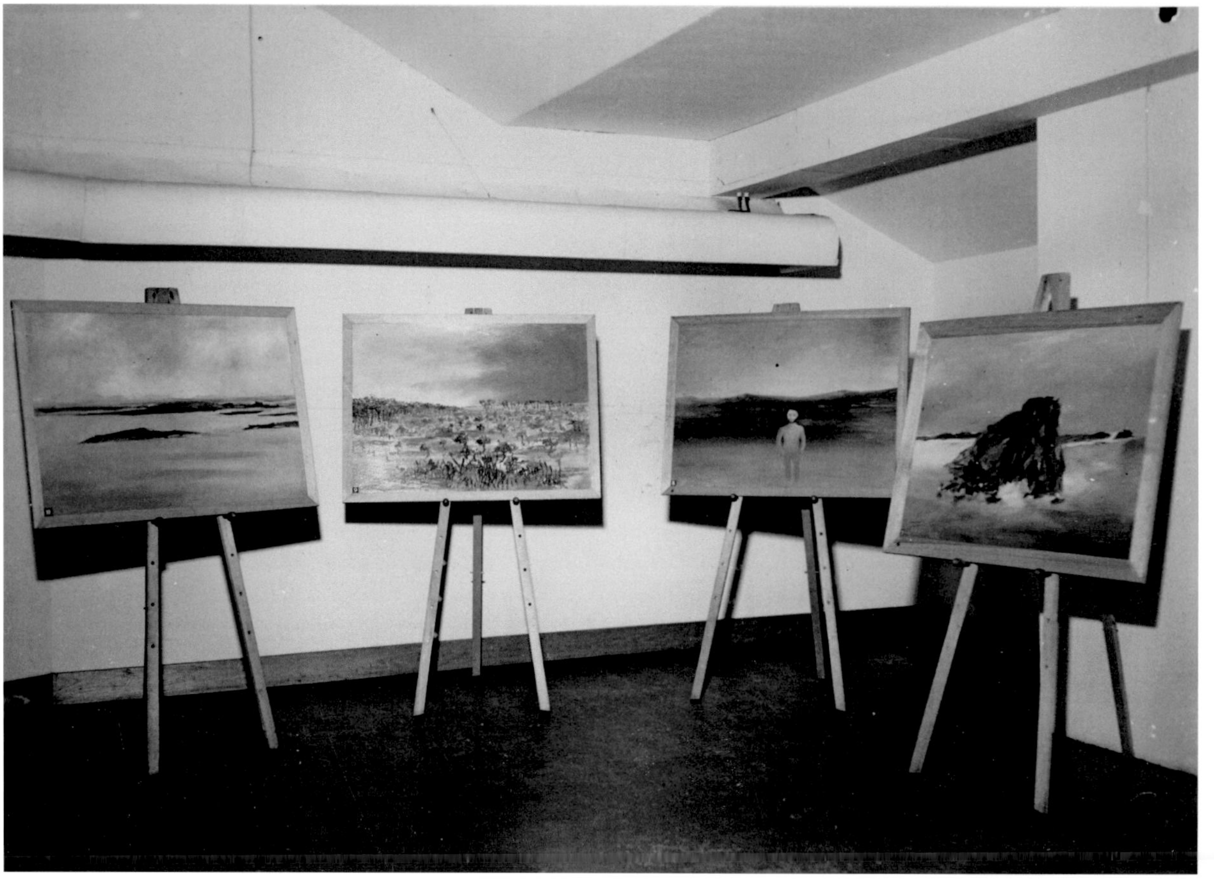

FIGURE 63
Exhibition of four of Sidney Nolan's Fraser Island paintings at Moreton Galleries, Brisbane, February 17–28, 1948. Left to right: #10 *Hervey Bay*, #9 *Swamp*, #8 *Island*, #7 *Indian Head*. Silver gelatin photograph. Cooper Collection, Fryer Library, University of Queensland.

The legacy of Nolan's time in Queensland is the evolution of his iconic Mrs. Fraser series. Technical and archival examination of six Fraser Island paintings has revealed that the first stages of *Mrs. Fraser* can be placed to Fraser Island in 1947 based on commonalities with the other works, particularly the reused condition of the Masonite support and the preexisting layer of green paint. The composition is rendered in Ripolin, Nolan's medium for all his Fraser Island paintings, presumably used from the satchel brought from Melbourne. While *Mrs. Fraser* has been associated with the bitter relationship Nolan subsequently developed with Sunday Reed, close examination of the work, and of the photo of the work taken by Nolan in 1947, gives clear evidence that the painting, with figure, was created in Queensland. Although this time coincides with Nolan's decision to end his relationship with Sunday Reed, he had not developed the antipathy toward her that many authors, and later Nolan himself, have suggested lies at the heart of the animal-like pose of the Mrs. Fraser figure.

When an amnesty for army deserters was declared in 1948, Nolan applied for and received a dishonorable discharge. This enabled him to exhibit openly—his Fraser Island paintings in Brisbane, and in April of that year the Kelly paintings in Melbourne. Over the next five years Nolan, Cynthia, and Jinx traveled extensively through central, northern, and western Australia. Nolan had a number of successful Sydney exhibitions with subjects derived from these travels and from photographs taken during his 1947 Queensland stay. After poor reviews in Melbourne and Brisbane of both the Fraser Island and Kelly exhibitions, it was Sydney that finally embraced Nolan's vision of place and people. Financial and critical success enabled him to embark on the next stage of his ever-changing repertoire of paint media and subjects.

Ripolin and Zinc Soaps

Ripolin is notable for being primarily formulated with the white pigment zinc oxide. The presence of zinc white in oil-based paints became increasingly common from the late nineteenth century, initially as a less toxic alternative to traditional lead white and later used in combination with titanium white to improve paint film properties. Its good suspension in oil transformed the viability of early ready-mixed house paint production. Although zinc white was consciously sought by some artists for its resistance to yellowing and blackening, it lacked the hiding power of both lead and titanium whites; it was therefore most frequently used in combination with other white pigments and in colored formulations. The use of zinc white without additions of other whites in the Ripolin colors prior to the 1950s is unusual and probably part of the attraction to Nolan, who particularly liked working with somewhat transparent paints.

Unfortunately, the reactivity of zinc oxide in oil is now associated with a variety of deterioration phenomena in paintings. Zinc soaps formed by reaction of the pigment with fatty acids derived from the oil medium have been implicated in formation of disfiguring lumps and textural change, insoluble surface deposits, and loss of adhesion and strength between or within paint layers. Paints with any amount of zinc oxide and a source of fatty acids appear vulnerable to this problem. Interestingly, though, Ripolin paints seem less prone to zinc soap–related degradation than many other zinc oxide–containing paints. When examples of deterioration with Ripolin have been reported, the paintings typically contain combinations with other paints or fatty acid sources (Townsend, King, and Ormsby 2016).

When zinc oxide is mixed with a drying oil, bonds rapidly form between zinc ions in the pigment and carboxylate groups in the oil. This interaction is evident in FTIR spectra of the paint as a broad vibration peak characteristically centered between 1560 and 1620 cm^{-1}. This broad peak is typically present in spectra of Ripolin paint samples taken from paintings by Nolan (fig. 64a). The zinc carboxylates do not in themselves appear problematic, and current thinking suggests the broad vibration reflects zinc ions paired with carboxylate groups still bound to the glycerol backbone of the oil (Hermans et al. 2015). In this form the zinc carboxylates remain connected and distributed throughout the polymerized oil network. If, however, there is a source of free fatty acids or the paint is exposed to conditions favoring hydrolysis of glycerol ester bonds, there is scope for untethered zinc carboxylates—especially zinc soaps of the long-chain saturated fatty acids, zinc stearate and zinc palmitate—to form. These mobile phases have a tendency to migrate and aggregate to reach a thermodynamically stable state as closely packed crystalline structures. The crystalline arrangement is reflected in FTIR spectra with a characteristic sharp antisymmetric

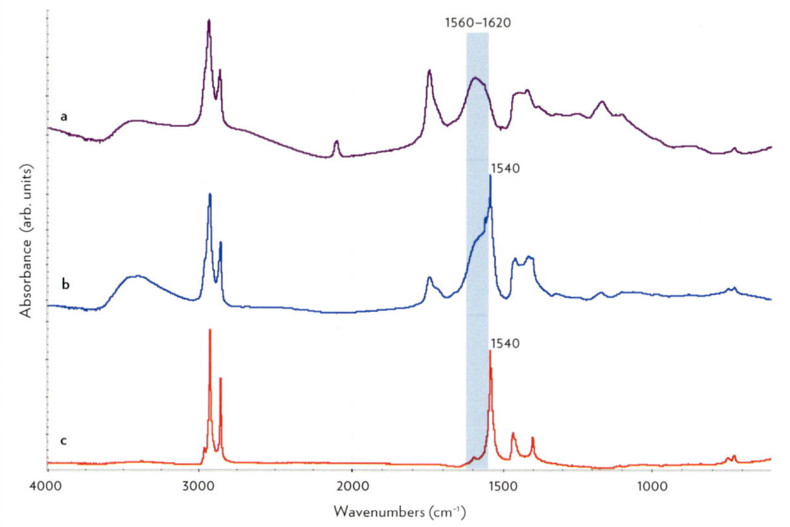

FIGURE 64
FTIR absorbance spectra of paint from (a) Sidney Nolan, *Mrs. Fraser* (1947; see fig. 53), dark blue paint (oil, zinc white, and Prussian blue); (b) zinc white oil ground layer from a painting dated 1980; (c) zinc stearate, QAGOMA synthesis.

carboxylate vibration at 1540 cm^{-1} (fig. 64b) similar to that observed in FTIR spectra of pure zinc stearate (fig. 64c). It is in this aggregated crystalline state that zinc soap–related deterioration is typically observed. The transition from network-bound zinc carboxylate to crystalline zinc soap does not consistently occur in Nolan's Ripolin paintings, and structural issues such as paint delamination are rarely reported.

Ripolin paints traded on their reputation as high-quality enamel paint. Nolan is one of many artists who went to great lengths to obtain Ripolin for its desirable optical and working properties, and these benefits appear to now also extend to superior stability. Careful and precise formulation of individual paints and the use of a heat-bodied oil is likely to have contributed to their relative stability (Carter et al. 2019).

The specific grade of zinc oxide pigment used in Ripolin paint formulations is another variable that may influence its reactivity. Zinc oxide particle properties determined by production method were central to mid-twentieth-century paint industry research aiming to improve stability of zinc oxide–based oil paints (Osmond 2012; Rogala 2019). Macro properties including particle size and shape have obvious influences on the reactive surface area of pigment stocks, but it has recently become possible to characterize sub-micron-scale properties of archival zinc white pigment samples using photoluminescence microimaging techniques (Bertrand et al. 2013). While the reactivity of zinc oxide in oil mediums complicates retrospective analysis of pigments used in historic paints and paintings, there is potential to probe the precise properties of zinc oxide present in Ripolin paints among Nolan's studio materials for possible insights. Further research into both zinc oxide pigment selection and the tailoring of oil media for individual Ripolin formulations may assist understanding of why these paints appear less susceptible to problematic zinc soap formation than many other zinc oxide–containing oil paints. The relative stability of paintings produced using Ripolin remains an intriguing observation.

5 Sydney and London Success: 1948–1982

When Nolan married Cynthia Reed (1908–1976), she was an author with two published books. Over the following three decades she would write five more in which she chronicled her travels with Nolan in Australia, North America, Africa, New Guinea, and China. Nolan, for his part, turned to new painting subjects during the five years he remained in Sydney, before he, Cynthia, and Jinx moved to London in 1953. First, outback Queensland, then northern and central Australia, and in 1953 animal skeletons, inspired by a trip to the Northern Territory, which was experiencing an extreme drought.

Nolan's first Sydney solo exhibition was at the David Jones Art Gallery, opening on March 8, 1949, with paintings of Queensland rural towns, birdlife, and people. It was an exceptional success. Sydneysiders bought the paintings, and the exhibition established Nolan as an emergent modern artist with a powerful vision of the Australian landscape. The income generated from the sales enabled Nolan, Cynthia, and Jinx to travel further through central and northern Australia in 1949, where Nolan built a catalogue of visual references for future paintings.

Over the following four years Nolan was to have six successful solo Sydney exhibitions. He continued to frame the works in native timbers of various types reminiscent of the tulip oak frames in which the Kelly paintings were first exhibited and of the native pine frames used with the Fraser Island paintings. One work, photographed along with its three-part beveled frame, the wood's distinctive dark knots harmonizing with the boab nuts in the painting, made the July 25, 1950, cover of the *Australian Women's Weekly* (fig. 65). This departure from the usual covers of the popular magazine indicated just how quickly Nolan's paintings were becoming synonymous with fashionable taste and contemporary design. Nolan's exhibitions also continued to utilize unusual display techniques. The drought paintings shown at the David Jones Gallery in 1953 were suspended from the ceiling such that they floated away from the wall (fig. 66).

Following the March 1949 exhibition, the Art Gallery of NSW purchased *Pretty Polly Mine* (1948; fig. 67), the first of Nolan's paintings acquired for a public collection.[1] At the time of the purchase Nolan was asked to fill out a standard form given to all living artists whose works were acquired by the gallery. In addition to biographical information such as training and exhibition history, there was a section on "mediums employed etc.," under which Nolan wrote: "Ripolin, oil, watercolour, lacquer, etching, glass, monotypes."[2] Thus Nolan acknowledged for the first (and only time) that lacquers, presumably nitrocellulose, had formed a key part of his technical repertoire.

Pretty Polly Mine is typical in technique for this period. Gloss Ripolin paint has been thickly applied in the sky, creating a ridge of media along the horizon. This effect may have been caused by Nolan moving the painting upright before the paint had dried

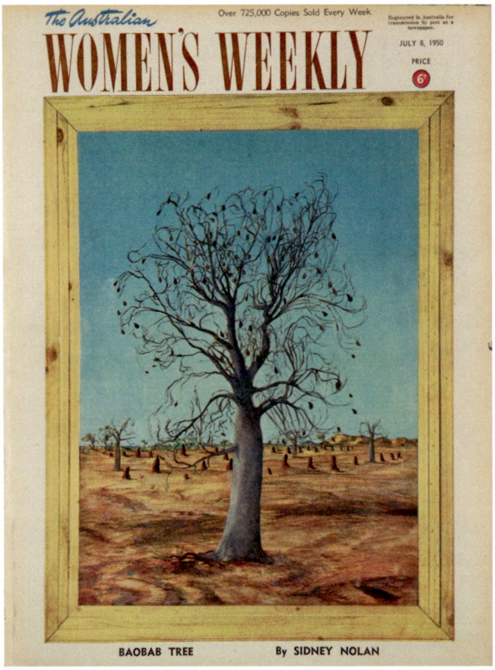

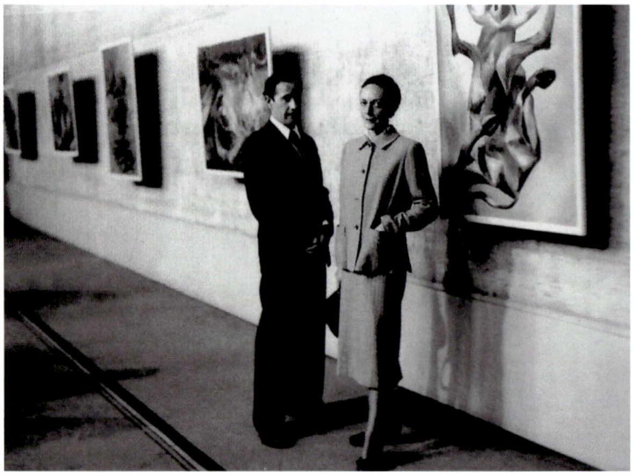

FIGURE 65
Sidney Nolan, *Baobab Tree*, 1950. Cover of *Australian Women's Weekly*, July 25, 1950. National Library of Australia.

FIGURE 66
K. Redshaw, Sidney and Cynthia Nolan at David Jones' Art Gallery, Sydney. *Sydney Morning Herald*, July 29, 1953.

through its full thickness or working with the painting propped up, as shown in photographs of him in his Sydney studio at his home at Wahroonga (fig. 68). In contrast to the thickly painted sky, the landscape has been worked thinly with scumbled Ripolin, leaving exposed in many areas the unprimed hardboard support. In this instance, the support is not a dark-colored Masonite board made from Australian hardwoods but a pale board made from softwood. Perhaps it was the board's inherent color that inspired Nolan to leave it unpainted between swatches of color in the foreground. However, over time the exposed areas of the board have darkened, possibly through contact with moisture or solvents; as a result, there is now some discordance between the colored areas and those left unpainted.

In 1950 Nolan was preparing for a David Jones exhibition to be held in April while working on watercolors for a Macquarie gallery exhibition and a series of paintings on the subject of the nineteenth-century central Australian explorers Burke and Wills for exhibition in June. This prolific and frantic production, in which large numbers of paintings were produced in small timeframes between travels, became Nolan's preferred way of working, but one he sought, at least in this early period, to keep to himself. As he put it, "One has to learn to sprint while giving the appearance of walking."[3]

While painting the pieces inspired by his trips to Queensland and central Australia, Nolan's palette underwent a dramatic shift. The landscape itself probably dictated a move toward strong earth tones. One critic noted in viewing Nolan's paintings on exhibition in 1950, "When you enter the gallery the blaze of reddish-brown hits you like a ton or two of real red earth" (quoted in G. Smith 2003, 23). An undated note by Nolan describes the Ripolin colors he used for these paintings; apart from the Blue 17,[4] they relate well to the cans found in his Wahroonga studio: "Undercoating white (no. 1), canary (no. 4), red (no. 16), blue (no. 17), ochre (no. 56) and black (no. 1105)" (G. Smith 2003, 22). In a letter of July 4, 1957, he stated in relation to the central Australian paintings the reasons for his choice of Ripolin: "Well I've used Ripolin on Masonite because I found out that in order to reproduce this light, this very intense bright light

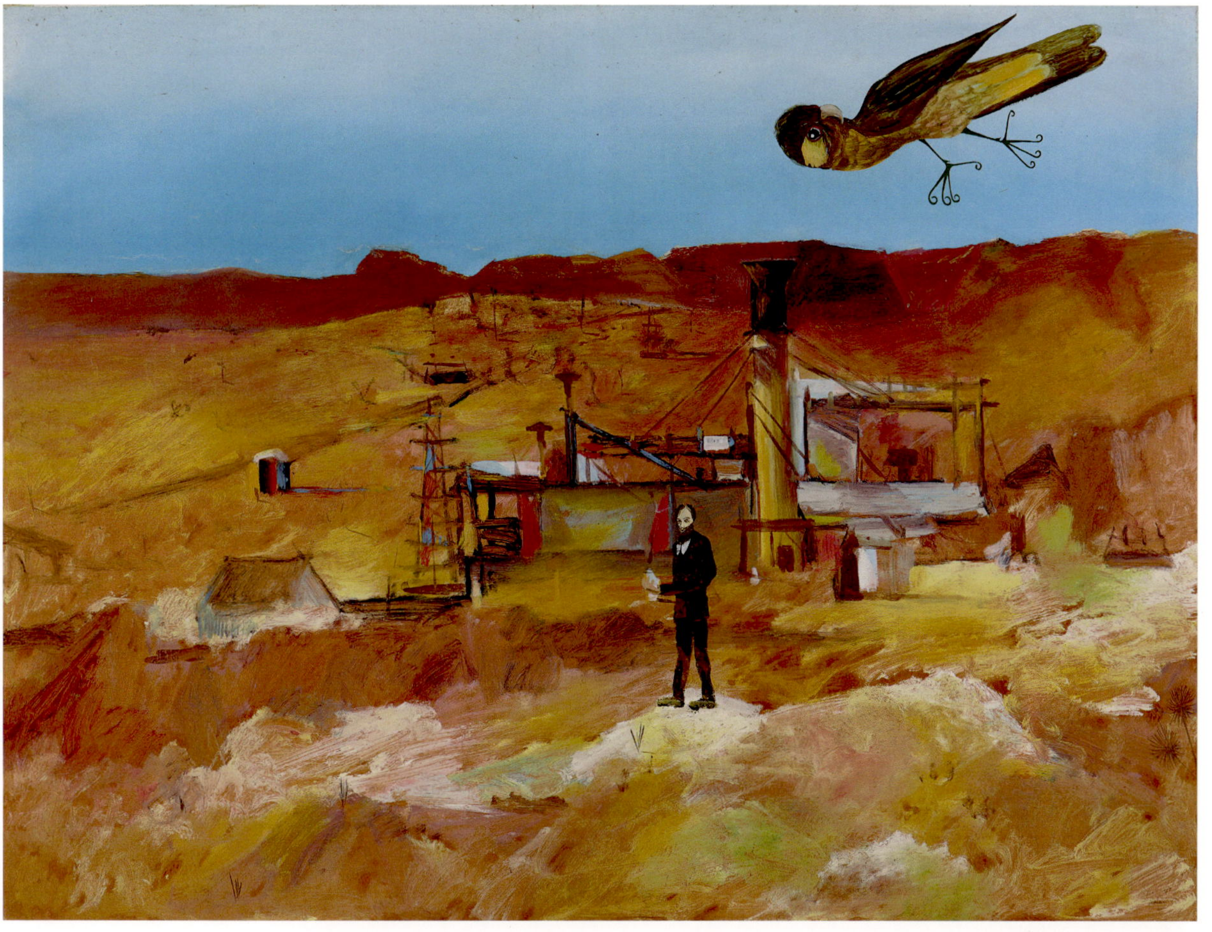

FIGURE 67
Sidney Nolan, *Pretty Polly Mine*, 1948. Ripolin oil enamel on hardboard, 91.0 × 122.2 cm (35⅞ × 48¼ in.). Art Gallery of NSW. Purchased 1949.

FIGURE 68
A. L. Frazer, Nolan painting *Unnamed Ridge* in front room of Wahroonga house, 1949. Silver gelatin photograph, 15.8 × 21.0 cm (6¼ × 8¼ in.). Cooper Collection, Fryer Library, University of Queensland.

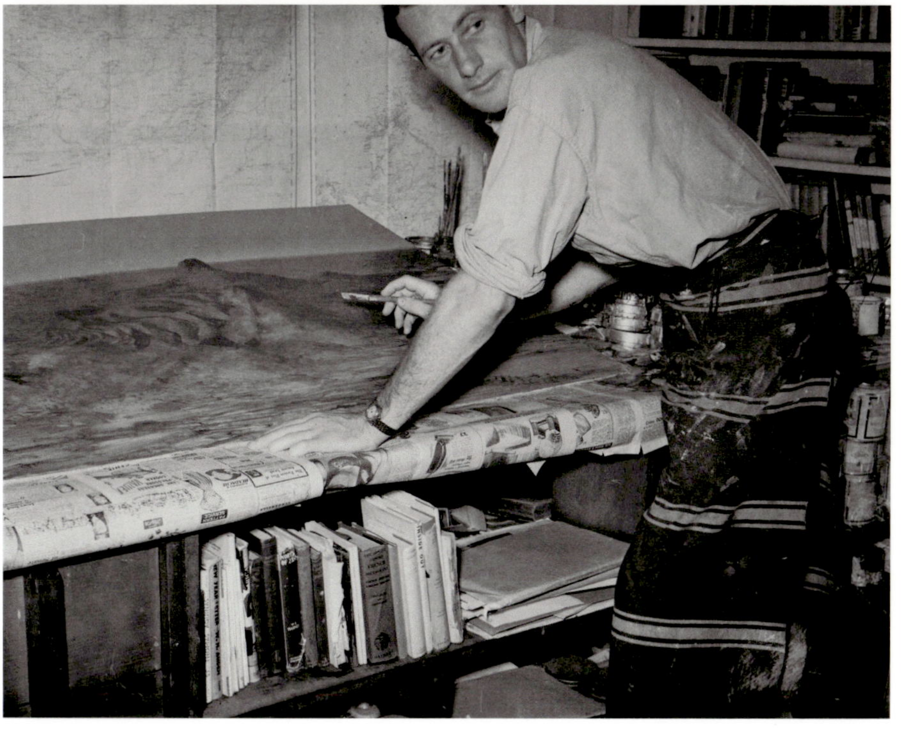

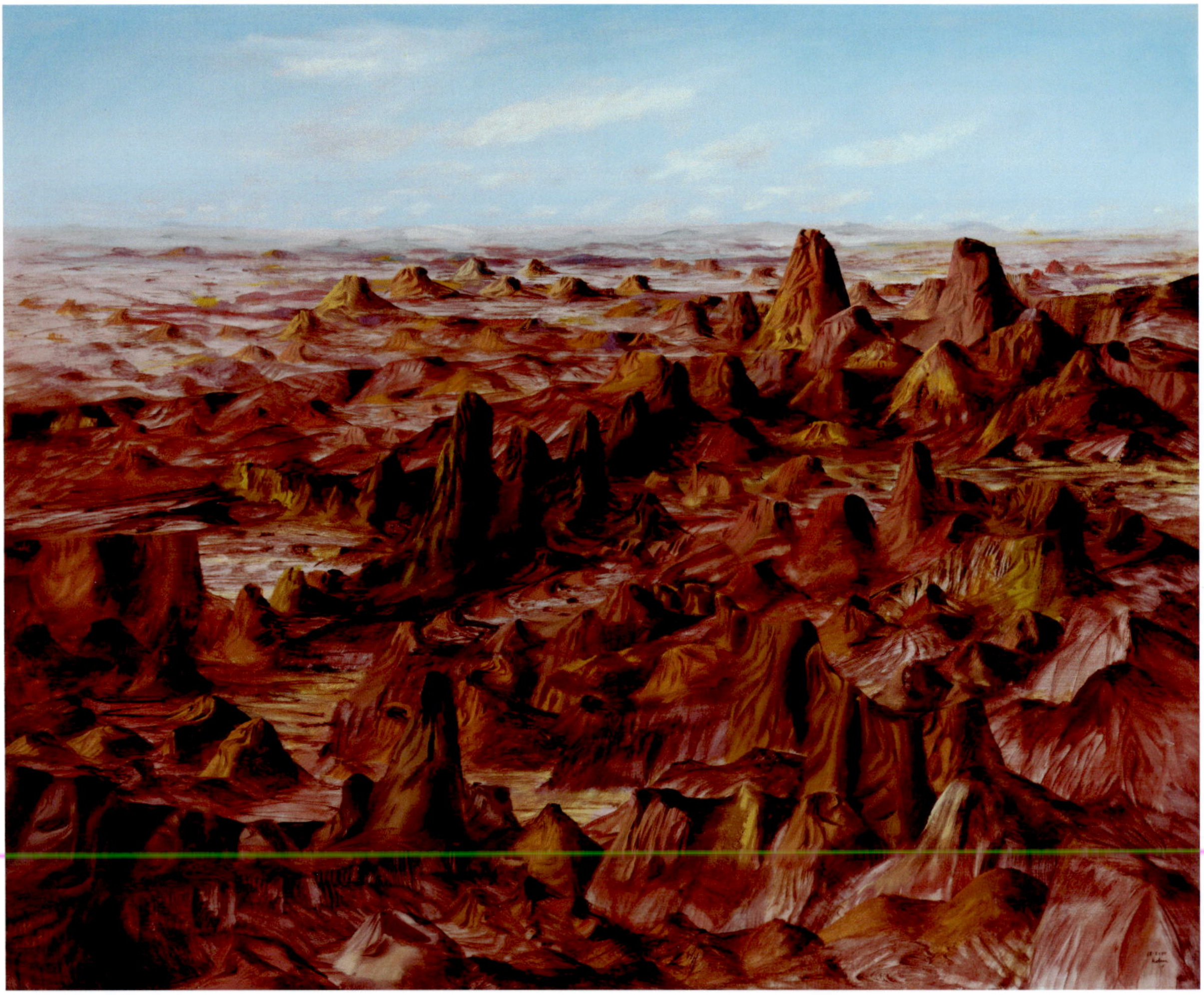

FIGURE 69
Sidney Nolan, *Central Australia*, 1950. Ripolin oil enamel and oil on hardboard, 122.0 × 152.5 cm (48 × 60 in.). Art Gallery of NSW. Purchased with funds provided by the Nelson Meers Foundation 2004.

which was flooding over everything, one had to adopt a different sort of palette from the conventional oil palette one and Ripolin on Masonite seemed to bring about this translucent sky and it set up the effect I wanted" (G. Smith 2003, 22).

It is unlikely, however, that Ripolin alone provided the strong red and brown tones in the Burke and Wills paintings from 1949 (see fig. 73) and the central Australian landscapes from 1950 (fig. 69). The media descriptions for many of these paintings include oil-based ochres, suggesting the addition of paints other than Ripolin.[5] Clues to Nolan's palette may lie in the contents of his Wahroonga studio.

The Wahroonga Studio

The house at Wahroonga was a modest weatherboard cottage with little room for a working studio for Nolan. Initially, he painted in the front room, as he wrote to Albert Tucker on January 2, 1950: "We were going to have a studio built out in the garden but it is all far too expensive, & all I require really is a big flat table & the fluorescent light to step up the sometimes dim daylight in the front room" (quoted in G. Smith 2003, 21). This arrangement must not have been entirely suitable, however, and in 1952 Nolan and Cynthia bought a house scheduled for demolition and set about dismantling it, then used the salvaged materials to build a garden studio. Nolan was at the same time considering a move to Britain, encouraged by the English art critic Kenneth Clark. Clark had seen Nolan's entries in the Australian landscape Wynn Prize exhibition in 1949 during a trip to Sydney and had sought out Nolan to see further works. As a result of Clark's support for his work and suggestions that Nolan should come to England, Nolan, Cynthia, and Jinx had traveled to Europe in 1950, and by 1952 Nolan was seeing the attractions of a longer stay. The completion of his new studio therefore was very shortly followed by a permanent move to London in 1953, leaving the studio and its contents in situ.

In 2007, while preparing the Wahroonga house for sale, Jinx Nolan packed the materials she considered most useful for study and gifted several boxes of paints, paper receipts, and small Masonite boards to the Art Gallery of NSW (see fig. 45) and a smaller box to the National Gallery of Victoria.[6] Some material was left behind, and in 2017 the new owners of the Wahroonga property documented the remaining items. In addition, it has emerged through study of Nolan's studio at The Rodd in the UK, his home for his final decade, that he took some materials with him to England.[7] Nolan's move to the UK and the closing of the Wahroonga studio in 1953 is a significant date marker, as it predates the availability and Nolan's use of the two new important plastic paints of the 1950s, polyvinyl acetate (PVA) and acrylic.

Although the Wahroonga studio contained a large amount of Ripolin, as discussed in previous chapters, other items suggest that Nolan engaged with a far greater variety of house paint media, pigments, and driers than has been previously considered (see appendix II). Some items are Australian Army supplies and may date from Nolan's wartime service. They include a bottle of department of defense–issued mosquito repellent, a ration tin of cocoa (now empty), and cans with army marking whose original purposes are unknown but that Nolan probably retained for use as mixing vessels. As to the remaining items, it is difficult to accurately date and assign them all to Nolan's artistic practice. Some materials, such as a tin of waterproofing bitumen paint, may have been used in the construction of the studio, though given Nolan's past use of unconventional materials, these cannot be completely dismissed from the list of materials potentially found on his artworks. Several cans of Kemtone™, a house paint based on glues and oils emulsified by soaps that was a popular interior wall paint prior to the development of synthetic alternatives, may also have been used either for the studio construction or as a priming medium for Masonite boards (or both). One can imagine that Kemtone's extremely matte and brilliant titanium white surface would have made an attractive priming for painting.

The Wahroonga studio also offers some clues regarding the change in palette of the central Australian paintings. A jar of dry yellow ochre pigment together with various labeled and unlabeled bottles of oil suggests that Nolan may have been mixing his own paint, and a stone slab with repurposed Vegemite jars containing thin translucent colors would also appear to be an arrangement of his hand-mixed paints (fig. 70). However, the most likely source for the oil-based ochre in the paint media descriptions of many of the central Australia and Burke and Wills paintings from 1949 and 1950 is suggested by a color chart for a paint tinter called Surtint™ (fig. 71) and one can of Liquid Stain Pale Chrome. The swatches on the Surtint chart are made from the tinter itself, and analysis of samples taken from the swatches demonstrates that most of the colors are a combination of chrome yellow (lead chromate) and iron oxides (synthetic) with drying oil (see appendix II, table 3). Prior to 1953, tinting machines were not available in decorating shops and house paints were available only in the manufacturer's range of premixed colors ("Paint Bar Where Colours Are Mixed" 1953, 22). Tinters such as Surtint,

FIGURE 70
Stone slab with recycled Vegemite jars used for mixing paints, Sidney Nolan Wahroonga studio. Artists' Materials Archive, Art Gallery of NSW.

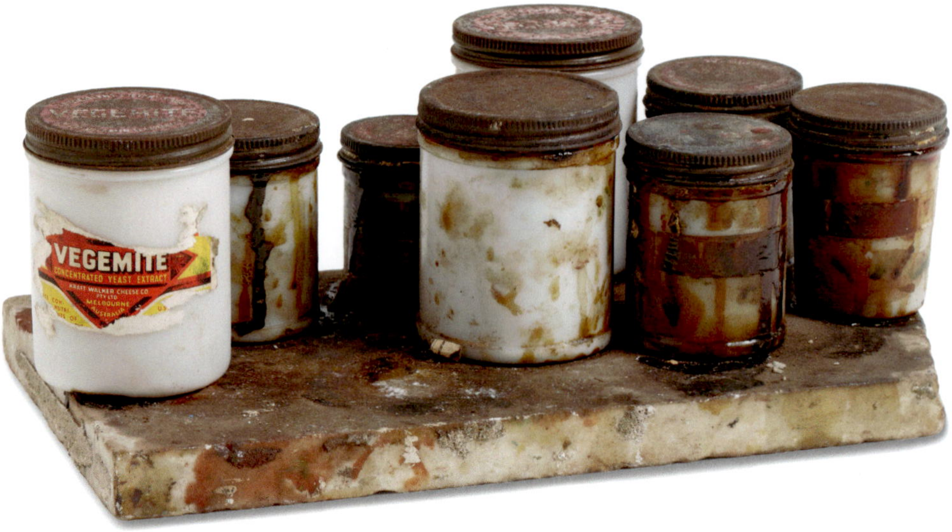

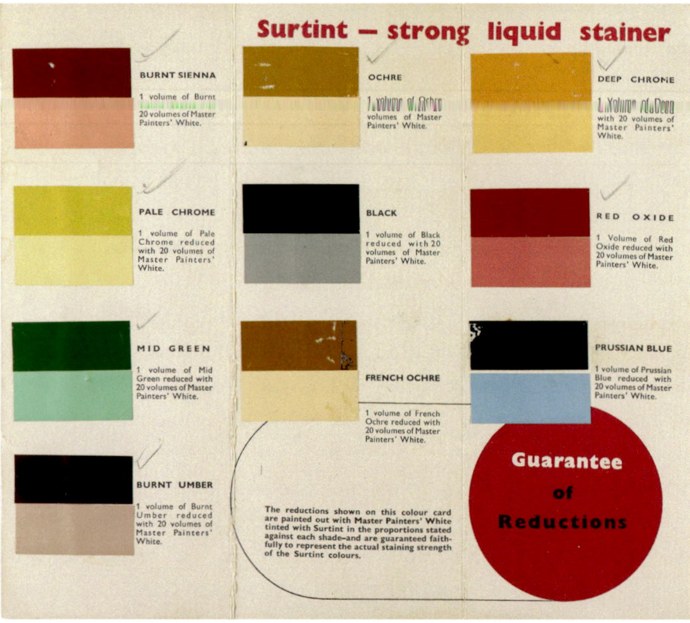

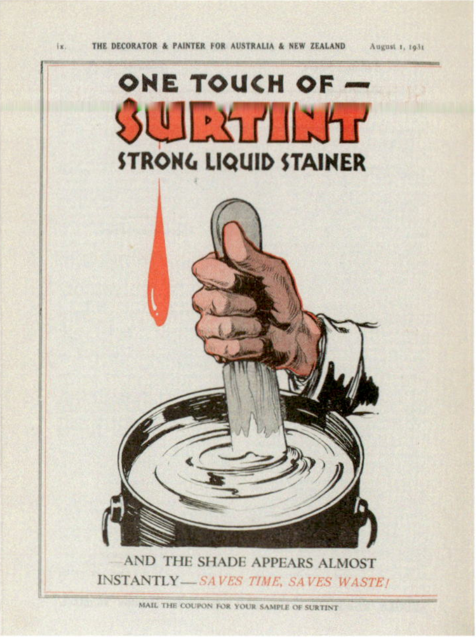

FIGURE 71 (left)
Lewis Berger & Sons, *Surtint Colour Chart*, from Sidney Nolan Wahroonga studio. Artists' Materials Archive, Art Gallery of NSW.

FIGURE 72 (right)
Surtint advertisement, *The Decorator & Painter for Australia & New Zealand*, August 1, 1931. Mitchell Library, State Library of New South Wales.

FIGURE 73
Sidney Nolan, *Burke and Wills Expedition, "Gray Sick,"* November 30, 1949. Ripolin oil enamel and oil on Masonite hardboard, 91.5 × 121.5 cm (36 × 47⅞ in.). Art Gallery of NSW.

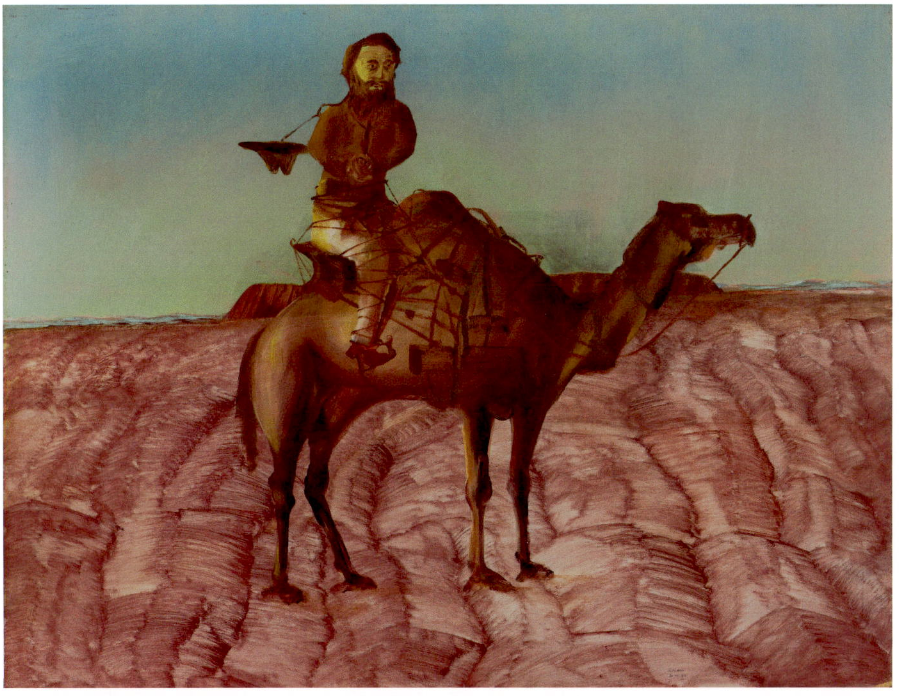

a highly pigmented paint, were used by trade painters to obtain custom colors by mixing with white paint to obtain the desired hue and tone (fig. 72). The undiluted tinters would have given Nolan useful options for obtaining strong colors for painting. The thinly scumbled foreground of the painting *Burke and Wills Expedition, "Gray Sick"* (1949; fig. 73) is a close match by hue and analysis to the red oxide color of the Surtint chart, though whether Nolan used Surtint specifically is impossible to say, given that red oxide and oil were common ingredients in both house and artist's paints.

Nolan's broad use of house paint media is further supported by several references in the Wahroonga studio material to a leading Sydney paint maker, Archibald Hatrick. Hatrick was originally an importer of New Zealand kauri resin for the paint trade. In 1934 he developed a relationship with an Austrian/American company called Beck-Koller, later renamed Reichhold, and began importing their synthetic resins ("A. C. Hatrick Limited" 1934, 56).[8] In 1937, Reichhold and A. C. Hatrick brokered a partnership to open a manufacturing plant for synthetic paint resins in Sydney. Production at Reichhold Chemicals Inc. Australia Pty. Ltd. began in 1940 at a plant in Botany, NSW ("Reichhold Chemicals Inc.: A New Australian Industry" 1940, 32–33). Reichhold Chemicals Australia did not market finished paint products directly to consumers, but rather provided the synthetic resins, pigments, driers, and technical expertise that local paint manufacturers needed to develop their own lines of paint. This enabled the large number of paint makers to keep current with overseas developments and compete with the larger companies that had direct licensing arrangements with American and British chemical companies. A. C. Hatrick was director of marketing and sales for the Reichhold factory.[9] A note in Nolan's hand on the Surtint chart reads "Ring Hatrick for English firm," suggesting that Nolan, in anticipation of his move to London, had sought information about potential suppliers of materials. Furthermore, a receipt for turpentine and stand oil purchased from A. C. Hatrick in 1952 has directions in Nolan's hand to the Reichhold factory and the name of the head of plant, Mr. Parche, suggesting that Nolan made at least one visit to the factory itself. Exactly what information or materials Nolan might have been seeking may be deduced from further notes by Nolan on the A. C. Hatrick receipt outlining a typical lead white paint formulation made from oil with the addition of metallic driers.

Driers

In the period prior to the development of plastic emulsion paints, driers were a critical component of house paints made from oil. House paints were typically fluid to allow brushing and flow of the paint, but the addition of a large proportion of oil medium made them slow to dry. Fast drying of the surface was critical for recoating and to reduce dirt pickup in the wet paint film. By the mid-twentieth century cobalt naphthenate driers were commonly used to encourage the rapid formation of a dry skin. However, ad hoc addition of driers to a paint mixture was not advised. Too much cobalt naphthenate and the paint surface would dry rapidly while the body continued to dry and shrink, causing wrinkling. Driers were therefore a specialist product not typically available to consumers. They were added to paints in very controlled and small quantities, often with the addition of other driers, called "through driers," which accelerated the drying of the paint body below the skin. The primary manufacturer of cobalt naphthenate in Australia in the 1950s was A. C. Hatrick at the Reichhold plant in Botany. It is likely that this was the material Nolan was seeking on his visit in 1952, as two bottles of cobalt naphthenate are present in the Wahroonga studio materials; they are well used, as drips on the outside of the bottles and the small amounts of residues left in the bottles attest.

Missing from the contents of the Wahroonga studio gifted to the Art Gallery of NSW are tubes of standard artist's oil paint. Photos from the current owners of the house and studio do show two tubes of oil paint that were left in situ during Jinx Nolan's cleanup in 2007 (personal communication 2017). Additionally, two small tubes of paint were recently found in Nolan's studio at The Rodd, in the case of materials and car parts Nolan took to the UK in 1950. However, the presence of just four small tubes of oil paint suggests that Nolan did not regularly use artist's paints before his move to London.

London

Nolan, Cynthia, and Jinx left Sydney in 1953, never to live permanently in Australia again. Nolan was based in London until Cynthia's death in 1976, and then in Herefordshire and the Welsh Marches through the 1980s until his own death in 1992. Nolan's endless curiosity and energy also propelled him to travel across the globe in an endless quest for new sights and experiences. He did not always paint while in the places he traveled to. The journeys were often difficult and not conducive to painting on site. In any case, such was not his usual practice. Typically, he would focus on his travels, then engage in intense and prolific painting sessions only upon his return. Still, he brought his art supplies along—just in case. "Sidney usually travels heavy, with full painting equipment, although he rarely paints while moving about. But what, he says, if he got stuck somewhere?" (C. Nolan 1967, 89).

After leaving Australia, Nolan extended his prolific painting career for four decades. This occurred alongside important advances in paint materials, most of which

Nolan took to as soon as they became available. For speed of action, Nolan needed media that dried quickly; his commitment to any type of paint or additive that assisted in this is therefore not surprising. Important indicators to his methods are found in the repository of materials left in Nolan's studio at The Rodd, an accumulation from his London years through his final decade in Wales. Other clues to Nolan's adoption of various materials are found in his diaries, in interviews, and in Cynthia Nolan's writing, but to date only limited analysis has been undertaken on his post-1953 paintings (see appendix I, table 2). Confusion continues, therefore, in the medium descriptions of Nolan's paintings through the 1960s to the 1990s. It is hoped that this study provides useful context and reference material for future analysis by owners and caretakers of Nolan's artworks post-1953, an area of little detailed investigation to date.

Polyvinyl Acetate (PVA)

In 1957, Nolan began working with the first plastic emulsion binder, available since the early 1950s. This was a plastic glue, made from a vinyl acetate polymer that, when mixed with water, created a versatile, quick-drying binder. The emulsification process[10] gave the glue its distinctive white appearance when liquid, and as a result it is commonly known as "white glue."[11] When dry, it became almost clear and no longer soluble in water, thus it is permanent and plastic. PVA was quickly adopted by the house paint industry as a binder for paint that could be easily cleaned out of rollers and brushes but, after drying, formed a durable, washable surface.

According to Wendy Whiteley, wife of Australian artist Brett Whiteley (1939–1992), PVA was everywhere in London in the 1960s. "PVA was the answer to everything, wasn't it.... Everybody used it. Sculptors used it, everybody used it. It just glued anything together and gave a glaze and you could mix it...mix up powder pigments. Worked terrifically" (Whiteley 2007, 113). When Nolan first started using PVA in 1957, however, only a handful of artists were aware of its potential as a paint binder that could be used mixed with pigments. These included the Italians Alberto Burri (1915–1995) and Lucia Fontana (1889–1968). Both artists used the PVA principally for its adhesion qualities, sticking collage and glass fragments to painting surfaces, though analysis of their work suggests that they may have also been making colored paint layers by mixing pigments and dyes into the PVA (Gottschaller 2012; Braun and Stringari 2015). In Italy, the brand of PVA available was called Vinavil®, and it was apparently from Burri via Albert Tucker that Nolan received information of this new plastic material. "I put him onto the polyvinyl acetate," Tucker said in 1979. "He dropped all the medium he was using, his Ripolin and so on. When he went out—he is a more organised man than I am—he quickly tracked down one of these firms and got this in. He threw out all his other stuff" (Gleeson 1979).

Exactly what brand of PVA Nolan was using in 1957 is not clear. Although the shelves at The Rodd have many large containers of polyvinyl acetate, they are branded Reeves and Art Spectrum, which didn't supply PVA to artists until about 1964.[12] Rowney marketed a range of artist's pigmented PVA paints in 1966, though no such colors were found in the studio; however, Nolan's personal documents do contain a color chart for these paints, suggesting that he knew of them and took an interest in them.[13]

In 1957 Nolan used PVA as a binder, mixing his own colors into the white glue. Initially he made both thick and liquid mixtures, depending on the type and quantity of pigments used. A large painting on canvas, later mounted onto chipboard, *The Galaxy*

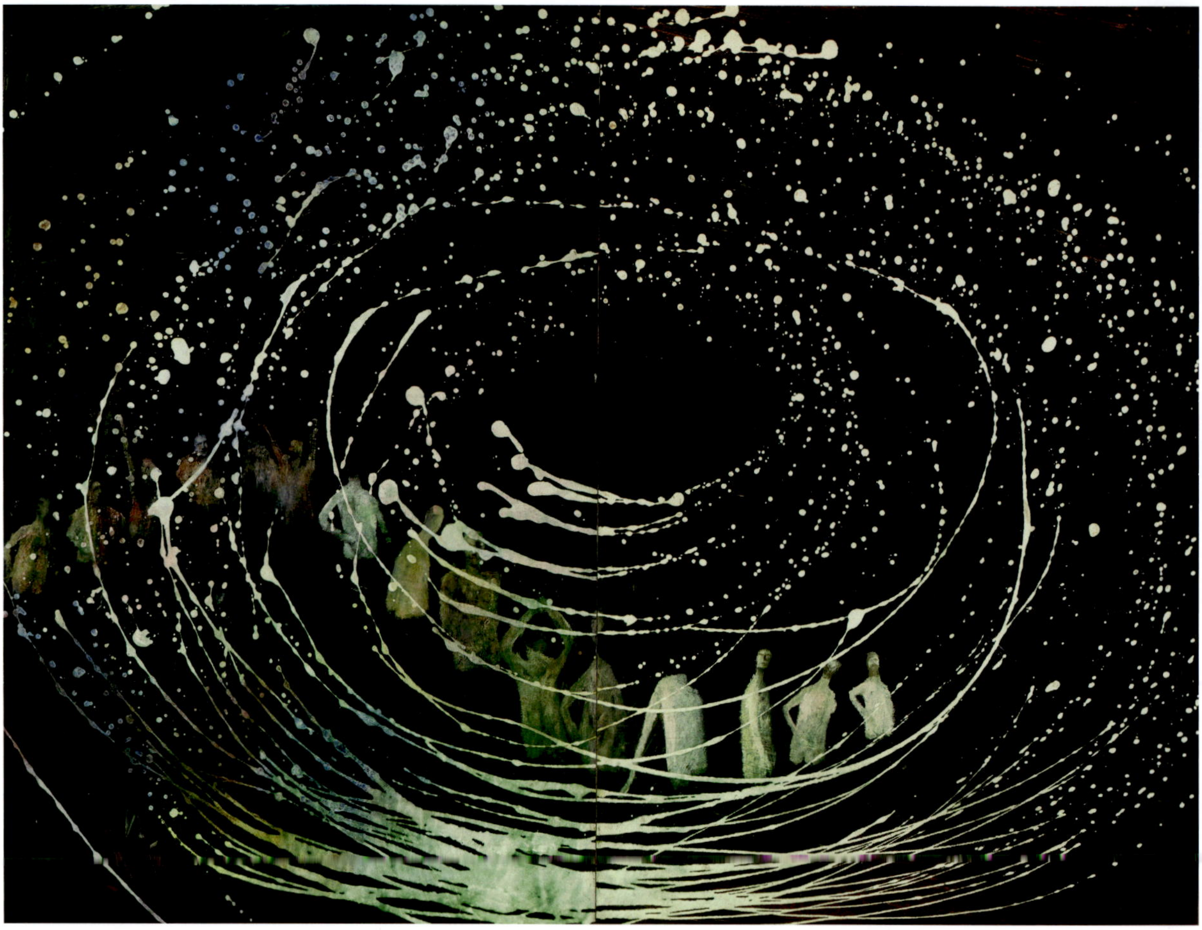

FIGURE 74
Sidney Nolan, *The Galaxy*, 1957–58. Polyvinyl acetate on canvas, 193.0 × 256.0 cm (76 × 100¾ in.). Art Gallery of NSW.

(1957–1958; fig. 74), is worked in dark layers of PVA scraped and wiped away to reveal faint figures, followed by an unusual white PVA paint dripped in trailing circular lines. The white pigment in the splattered paint accretions is poorly integrated, with crusty lumps that suggest it was roughly hand-mixed from dry pigment. Nolan spoke of the pleasure of working with PVA and the rush to complete a painting before the glue set:

> I used polyvinyl acetate because it was such a fast-drying medium that I had to work very quickly against it. It was like cooking a soufflé. There was a point at which it bubbled and hardened and you couldn't use it anymore; it was like lava. So one had a specified time for producing images, and this was quite exciting. . . . I like to work against the medium, so I keep changing. I think it's partly a result of working with different mediums as a boy in factories; all kinds of commercial synthetic substances so that now I have a taste for them. (Nolan, July 4, 1957, quoted in Crook and Learner 2000, 24)

By 1958 Nolan had developed a new technique for working with the PVA in which dark layers of green, brown, and Prussian blue were applied over white or colored backgrounds, then rapidly scraped back with squeegees and rags. *Leda and the Swan* (1958; fig. 75), for example, reveals figures and landscapes in the wiped-away shapes. In a photograph taken in Nolan's London studio in 1960, we see Nolan in the midst of the inevitably messy process (fig. 76). Cynthia described his work approach: "During the

FIGURE 75
Sidney Nolan, *Leda and the Swan*, 1958. Polyvinyl acetate on hardboard, 122.0 × 152.5 cm (48 × 60 in.). Art Gallery of NSW.

FIGURE 76
Gert Deutsch, Sidney Nolan in his St. John's Wood studio, London, 1960.

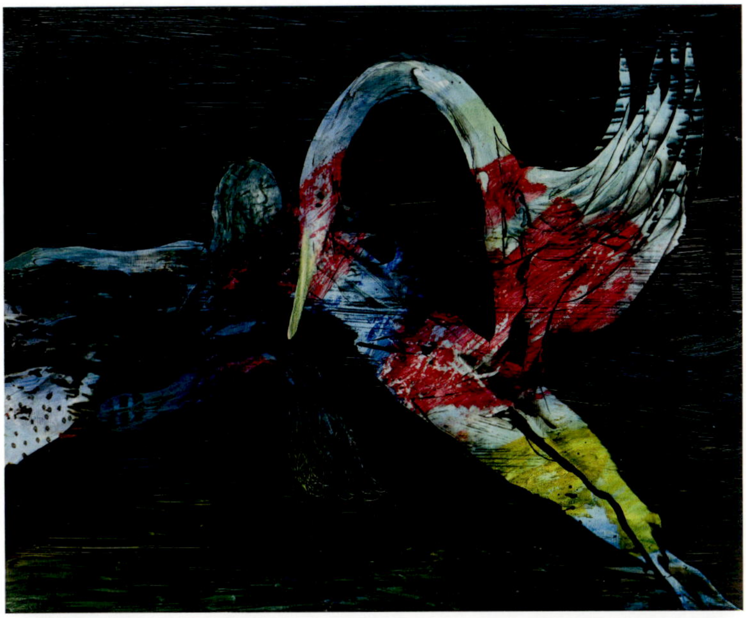

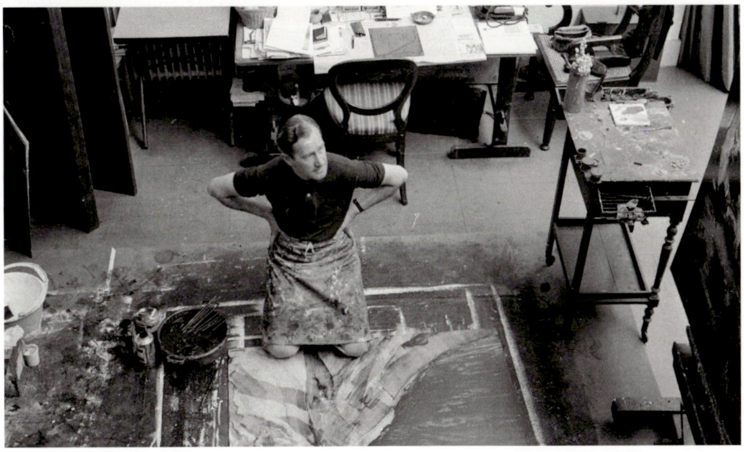

day he painted on the floor, first placing areas of colour on the prepared board, next sweeping on the polyvinyl acetate until the whole 4 × 5 feet area was thick with paint, then seizing a short-handled squeegee and slashing and wiping, cornering and circling like a skater, until another painting was completed" (C. Nolan 1967, 224). The layering and then revealing of forms through the absence of paint was a return to processes he had learned using spray paints and stencils for signs on glass in the 1930s.

In 1958 Nolan received a Harkness Fellowship, awarded by the Commonwealth Fund of New York City, to travel to the United States. Cynthia Nolan wrote of their American adventures in *Open Negative* (1967); for months they drove a convoluted zigzag across the country in a nine-seat station wagon. The cause of Cynthia's fatigue and illness during these travels was finally diagnosed as tuberculosis, and on their return to New York she was confined to a clinic for six months. The opening paragraphs of her book describe Nolan's and her arrival in New York by boat with an abundance of luggage, which included "six drums of polyvinyl acetate, fifty cans of paint, three boxes of dyes, five gallons of undercoat, ten packets of 10 × 12 inch art paper, two dozen 4 × 6 feet paintings on hardboard, one case of books. . . . Here I gave up counting" (C. Nolan

1967, 3). Why Nolan thought it necessary to transport such quantities of artist's materials to the commercial heartland of America is unfathomable, though it is consistent with his usual practice of traveling with copious supplies, "just in case."

The timing of Nolan's arrival in New York is interesting, for by 1958 an alternative plastic emulsion, made from acrylic rather than vinyl acetate, was becoming known to American artists (see sidebar on p. 92: Synthetic Emulsion Paints). Liquitex® was the first brand of artist's acrylic emulsion paint, available from 1954 but not used broadly until 1963 when it was reformulated in a thicker version (Crook and Learner 2000). Nolan must have been aware of acrylic, but it is a surprising absence in his own and Cynthia's records of his practice. Perhaps it gave no improvement to what Nolan was already achieving with PVA. That said, various Liquitex products were found in Nolan's studio at The Rodd, including a box of six colored Liquitex paints in small glass jars, plastic tubs of gesso, gloss polymer medium, matte medium, titanium white paint, modeling paste, and tubes of colors, gel mediums, and retarding medium, but only the acrylic gloss medium and varnish and several unused tubes of metallic paints predate the 1964 purchase of Liquitex from Permanent Pigments by Binney & Smith (Rajkowski 2013), suggesting that most of these acrylics date after Nolan's return from the US. In addition, a recycled plastic bottle hand-labeled "Primal AC634" (a type of acrylic polymer emulsion) and bottles of Art Spectrum acrylic emulsion, together with the Liquitex mediums, suggest that Nolan may have made his own paints by mixing pigments with the emulsion himself. Exactly how and when he incorporated acrylics into his practice, however, will not be known until further analysis is undertaken on his paintings.

A wooden case in the Rodd studio, with labels and inscriptions indicating it was used to transport materials back from New York on the *Queen Mary* in 1960, contains paper packets of dry pigments purchased from a New York color purveyor, Fezandie & Sperrle, and boxes of wax sheets. Located at Fulton Street on the Lower East Side, Fezandie & Sperrle was a well-known artist's materials supplier, associated particularly with the New York artist Jasper Johns (b. 1930), who purchased his encaustic beeswax medium from them through the 1950s (Stavitsky 1999).[14] Nolan, too, had been interested in wax as a paint medium as early as 1938–39, when he used wax boot polish on his painting *Head of Rimbaud* (see fig. 7). There are also several tins of shoe polish in the Rodd studio alongside large slabs of wax, boxes of red buffing wax, and small cakes of refined beeswax.

Nolan turned again to wax mediums and wax resist techniques during his time in the United States. Cynthia Nolan wrote about him painting after their long road journey across the country: "He now sat down to work, rapidly, in a mixture of watercolour paints, dyes and wax media," she wrote. "Sidney was making *sotto voce* remarks as he used his fingers, a brush, crayons, a rag, then held the page firmly, burnishing it with one of his best handkerchiefs" (C. Nolan 1967, 91). At this time, Nolan was also working with wax crayons and dyes on cards coated with a layer of smooth kaolin, into which the dyes soaked except where they were covered by wax markings.[15] One such work, *Landscape No. 27 (Convalescence)* (c. 1962; fig. 77), was likely crafted in this way. Made to illustrate a published collection of poems by Randolph Stow (1935–2010), *Outrider: Poems 1956–1962,* this work illustrated the poem "Landscapes" ("A crow cries: and the world unrolls like a blanket . . ."; Stow 1962, 38), but it also references the poem on the facing page, "Convalescence," the bubbles and striations in the paint evoking Cynthia Nolan's own illness and struggle with tuberculosis.

In both the PVA paintings on Masonite and these small kaolin-on-paper works done in New York, dyes rather than dry pigments are an essential part of the process. Mixing pigments into the fast-setting PVA, as he had done in *Galaxy,* was problematic,

FIGURE 77
Sidney Nolan, *Landscape No. 27 (Convalescence)*, 1962. Synthetic polymer paint on paper, 30.5 × 25.5 cm (12 × 10 in.). University of Western Australia Art Collection, Tom Collins Bequest Fund, 1962.

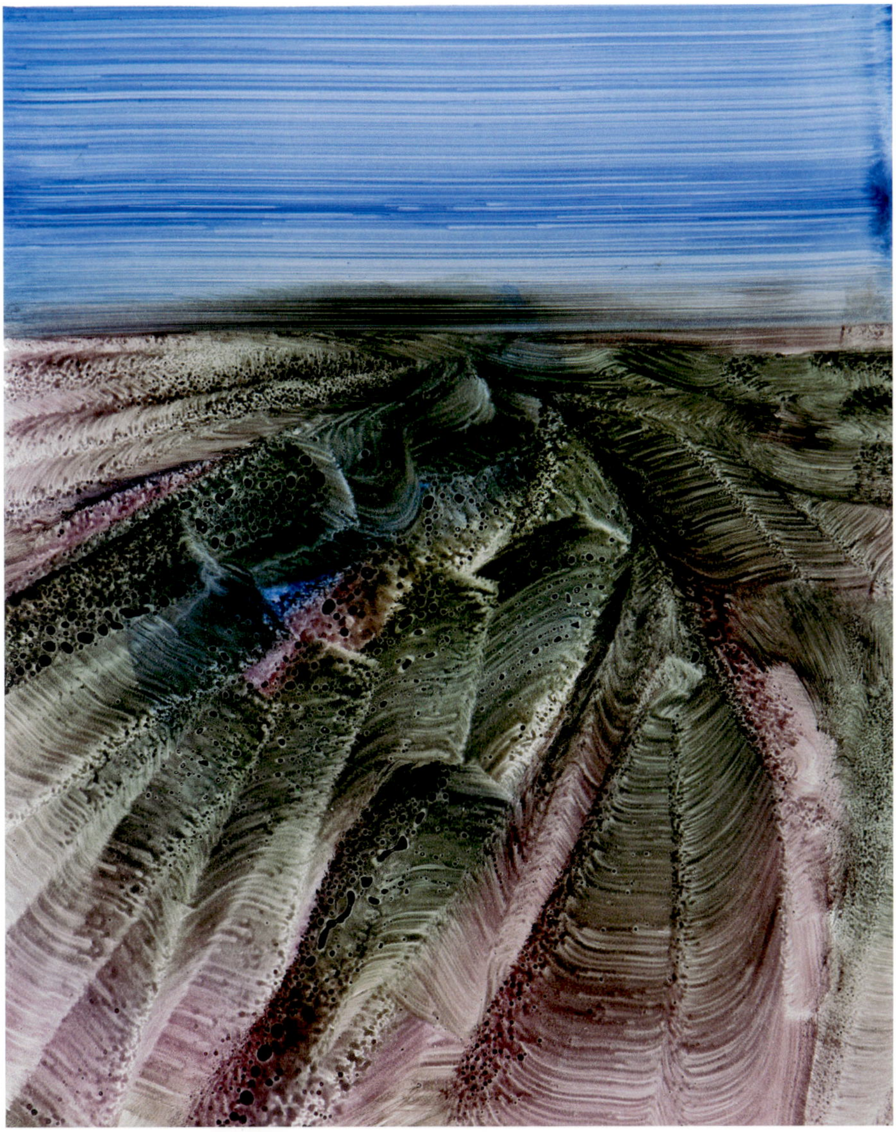

for the dry pigments were difficult to disperse quickly enough and tended to clump and form crusty paints. This may be why Nolan turned to the use of Helizarin® textile dyes.

Helizarin Dyes

The evidence in the Rodd studio of Nolan's first use of fabric dyes points to 1957, the same year he took up PVA. Among the materials Nolan brought back to London from New York in 1960 is a cardboard box containing the pigments purchased at Fezandie & Sperrle. However, the box's shipping label indicates that it originally contained Helizarin dyes which Nolan ordered from his home in Putney in December 1957[16] from Skilbeck Brothers, a dye and chemical company based in London since the 1650s (Dawe 1950).[17] The label is partly lost, but remnants of the names of the colors the shipment contained are consistent with a collection of small glass jars of Helizarin dyes still in the studio. Helizarin is a proprietary name given to a group of textile dyes and binders for fabric printing, manufactured by the German chemical company BASF. Helizarin colors are dyes made into pastes with dispersing agents. These powerful tinters would have integrated into the PVA emulsion quickly and without the miscibility issues of dry pigment particles that Nolan no doubt experienced. By the twentieth century, Britain's lead in the production of synthetic dyes had been superseded by German industrial chemistry, particularly of organic dyes derived from coal tar (see sidebar on p. 18: Coal-Tar Colors). The Helizarin dyes that Nolan purchased in 1957 from Skilbeck would have been imported from Germany.

Nolan's interest in fabric dyes for painting may have developed in part through his engagement with stage production. In 1961 he was commissioned to design and produce the sets and costumes for a ballet at the Royal Opera House, a restaging of Igor Stravinsky's *The Rite of Spring* (Daniels 2017). For the costumes, the dancers wore body stockings that Nolan painted using stencil handprints cut from newspapers; these stencils were pinned in place and Nolan blew paint through an aspirator onto the stockings while they were being worn (fig. 78). The use of stencils and aspirated paint again harks back to Nolan's work as a commercial artist spray-painting lettering, and in this instance is a direct reference to the paintings Nolan saw painted on the walls of Carnarvon Gorge, Queensland, in 1947.

A drum of Helizarin binder TS in the studio at The Rodd may also date from Nolan's work in *The Rite of Spring*. FTIR analysis of the drum's contents identified the pulpy and pale mass as an emulsion of polybutadiene acrylonitrile,[18] a synthetic rubber commonly known as nitrile. It was initially called Buna N when it was developed by the German conglomerate IG Farben Industries during the Second World War (Aftalion 2001, 148). After the war, IG Farben was disbanded into its smaller entities, including BASF, which gave the proprietary name Helizarin to its range of fabric dyes and binders. Whether Nolan used the nitrile rubber binder more broadly than on theater costumes but also as a paint medium is an intriguing idea and one worth considering in the further analysis of his paintings.

Another set of Helizarin colors are present in the studio in plastic containers labeled with the name of the distributor, Brico, the likely successor to Skilbeck Brothers, which closed in the 1970s (fig. 79). A full list of the Brico Helizarin dyes, with FTIR analytical results, is given in appendix II, table 5. It is a selective range of colors based on strong tinting pigments, including phthalocyanine blue and green and coal-tar dyes, azo reds and oranges and diarylamide and diarylide yellows. Exactly how Nolan used these dyes into the 1970s is not yet known. He must have mixed them with other

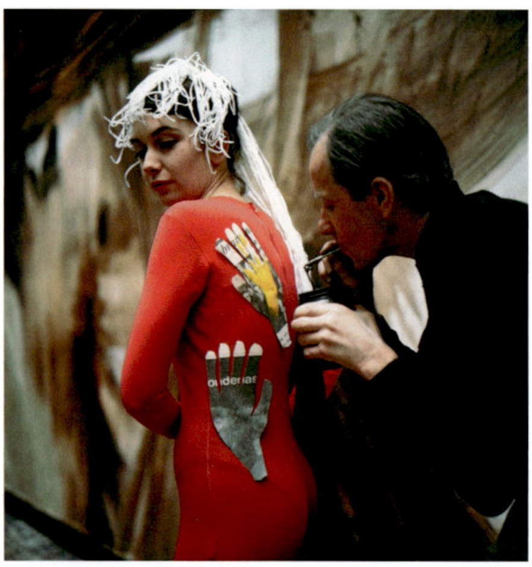

FIGURE 78
Zoë Dominic, Sidney Nolan working on the costumes for *The Rite of Spring*, 1962.

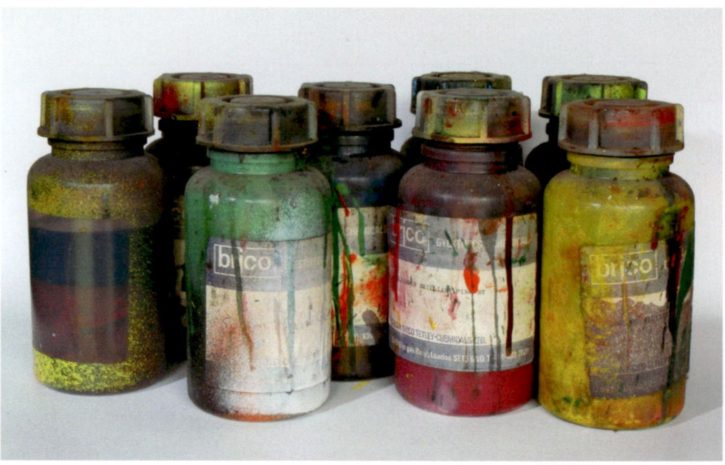

FIGURE 79
Brico Helizarin colors in Nolan's studio at The Rodd, 2017.

FIGURE 80
Philip Quirk, Sidney Nolan at his Ruthland studio, England, 1980.

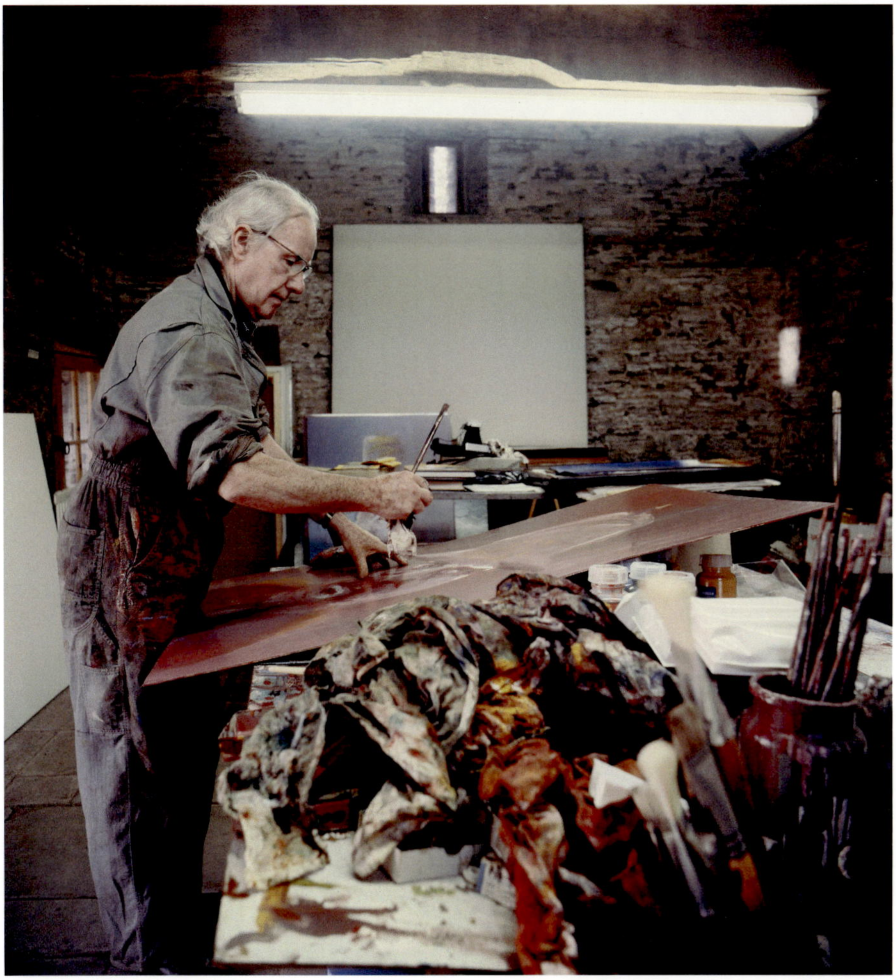

materials, as they have no binder as supplied, but when sampled in 2017 they stayed wet for weeks, drying slowly into a crumbly mass. Typically, dyes were mixed by the manufacturer with wetting agents for rapid dispersion into a binder, and FTIR analysis of one of the Brico colors, turquoise GT, suggested the presence of butylated urea formaldehyde, likely the dispersing agent. A 1980 photograph of Nolan painting at Ruthland, his Herefordshire home prior to The Rodd, shows the plastic jars of Brico Helizarin being used to prop up a painting in progress, but the photo does not indicate what paint binder Nolan might have been mixing with them (fig. 80).

Artist's Oil Paints and Alkyd Gel Medium

After returning from New York in 1960 and following a 1962 trip retracing the travels of his favorite poet, Arthur Rimbaud, through Kenya, Tanzania, and Ethiopia, Nolan embarked on a series of paintings of African animals and figures. Though continuing to paint on hardboard with white PVA priming layers, he now shifted his approach, adding colored layers with soft brushing and blending, a technique more associated with the use of artist's oil paints. Final touches of vegetation were applied with a small brush. Skies are densely painted and may be Ripolin enamel. This is an overlooked period of Nolan's work that warrants closer analytical study. Paintings of African subjects from

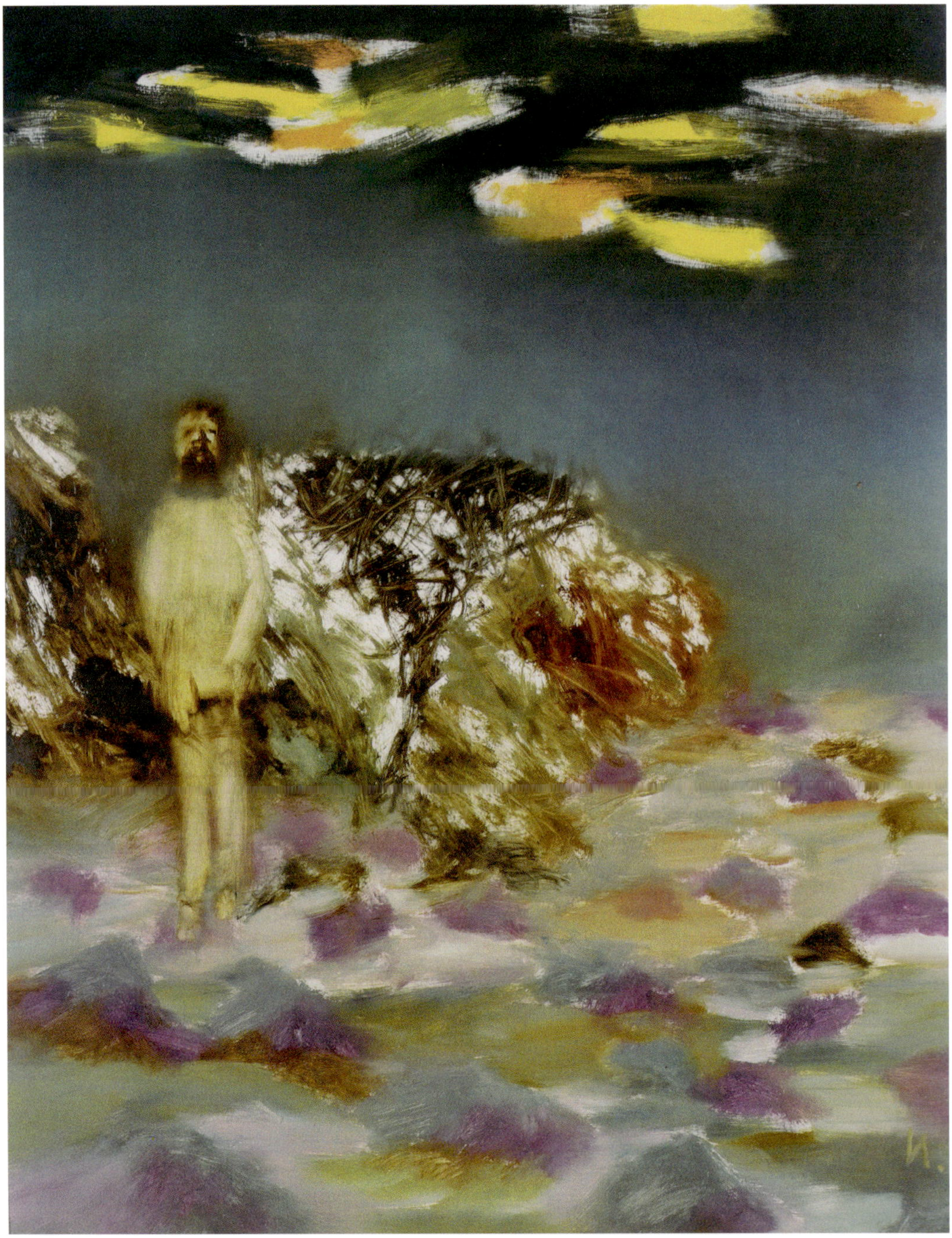

FIGURE 81
Sidney Nolan, *Rimbaud at Harar*, 1963. Oil on hardboard. 152.5 × 122.0 cm (60 × 48 in.). Private collection.

1963 such as *Rimbaud at Harar* (fig. 81) suggest that artist's oils may have finally become a medium of interest, though when they were used, paint layers were kept thin to maximize quick drying. Nolan and Cynthia bought a house on the Thames at Putney, with a spectacular view down the river to London. *Thames* (1962; fig. 82) offers a unique exception of a painting by Nolan of his adopted country: its softly brushed effects are unlike any other of Nolan's works in PVA paint, warranting analytical investigation.

The following year, however, a more satisfactory solution to the problem of slow-drying artist's oils presented itself to Nolan, spurring him into a frenzy of activity in which he produced large numbers of paintings in an incredibly short period of time.

FIGURE 82
Sidney Nolan, *Thames*, 1962. PVA on hardboard, 122.0 × 122.0 cm (48 × 48 in.). Private collection.

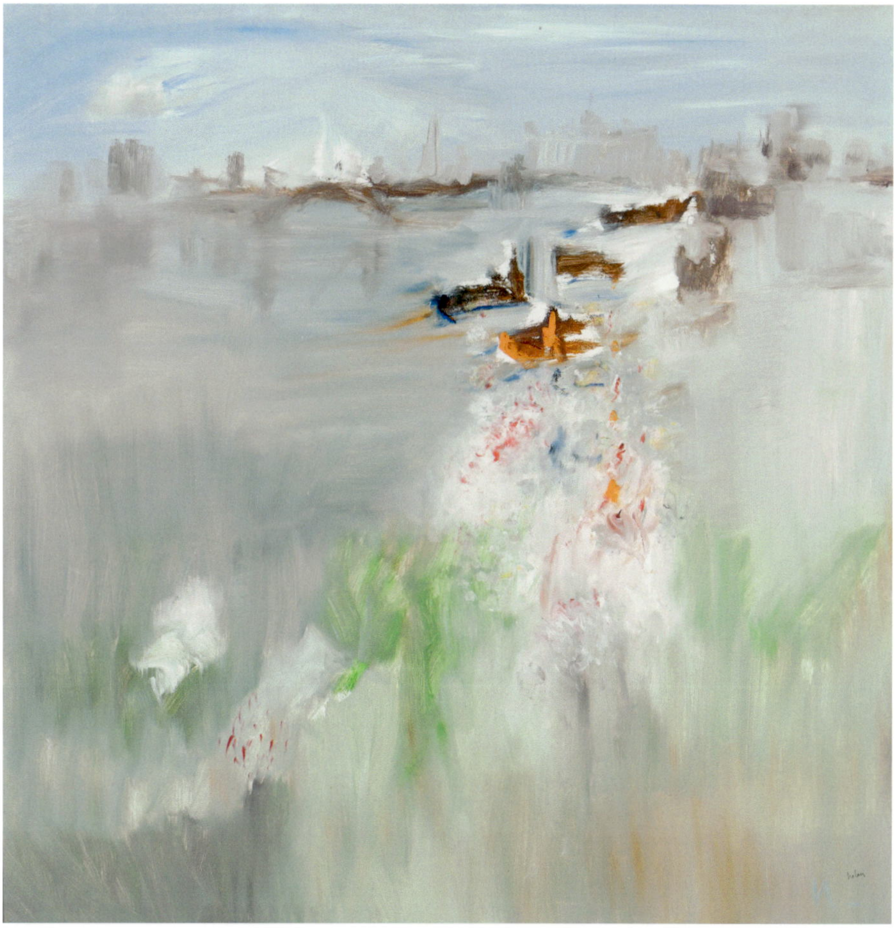

Released in 1963 by Rowney, this new medium, a clear thixotropic alkyd gel, allowed for speed drying of traditional artist's oil paints. Over 280 tubes of Rowney's alkyd gel medium in three generations of packaging in the Rodd studio attest to Nolan's affection for this material (fig. 83). Nolan developed his own way of using this material, in a series of paintings of Antarctica begun in 1964 after he had spent time on the ice. Described by his widow Mary Nolan in 2008 she said: "He squeezed it out on his canvas like that, just the gel, and then he put the paint on that he wanted and rubbed it with his hand, like that, and he rubbed and rubbed and rubbed; and he got the most amazing effects for sky and water and stuff like that" (Ormsby and Smithon 2008).

FIGURE 83
Boxes of Rowney's alkyd gel medium in Nolan's studio at The Rodd, 2017.

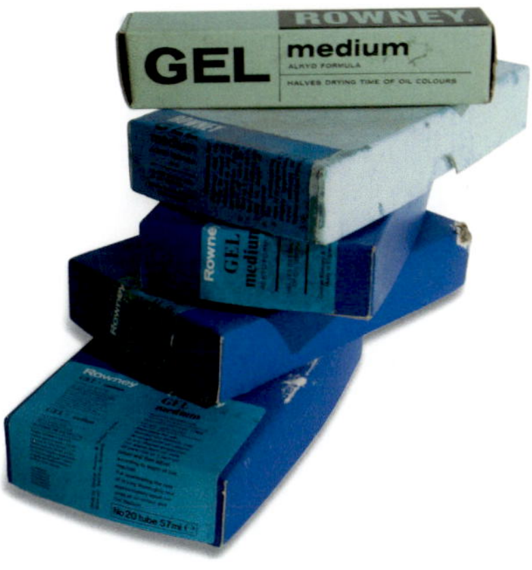

FIGURE 84
Sidney Nolan, *Icebergs McMurdo Sound, Antarctica*, 1964. Oil on hardboard, 122.0 × 122.0 cm (48 × 48 in.). Private collection.

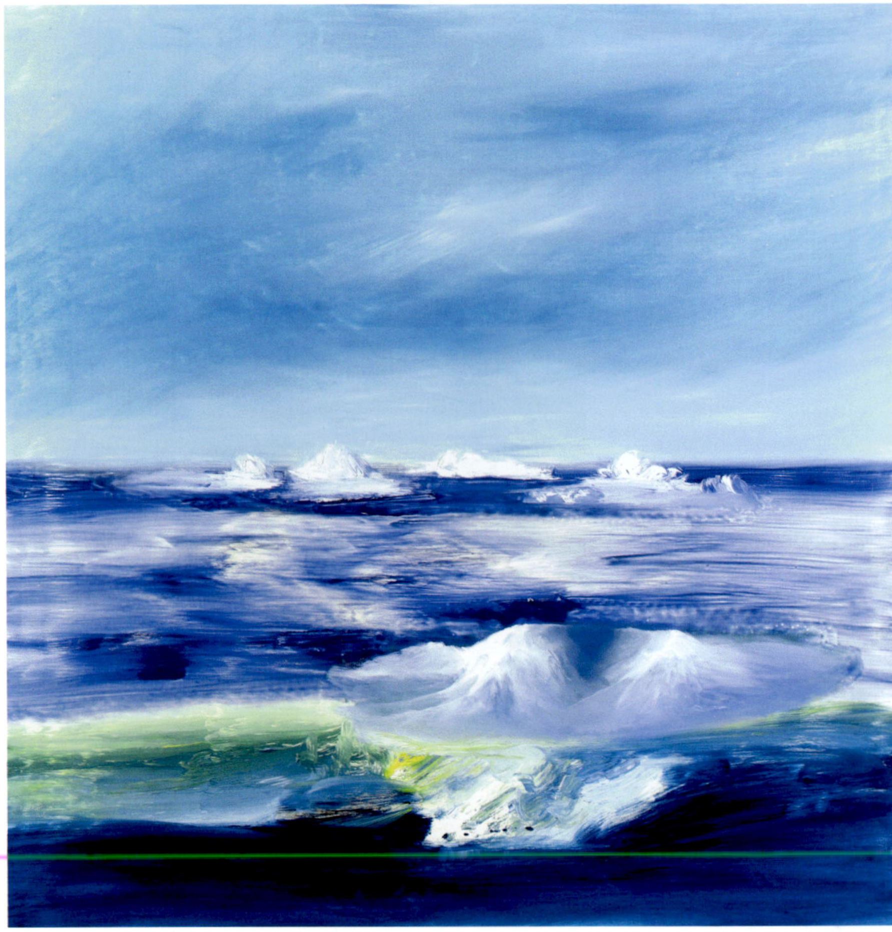

Along with an unusual transparency imparted by the gel, the evidence of finger painting is evident in the paint texture of many of these Antarctica paintings, such as *Icebergs McMurdo Sound, Antarctica* (1964; fig. 84). However, FTIR analysis of several samples from one painting, *Mt. Erebus* (1964), failed to positively identify the presence of alkyd on the work (see appendix I, table 2). This highlights the difficulty of analysis with microscopic-sized paint samples, which can be too readily interpreted as indicative of the whole of a painting—something particularly problematic in paintings with incompletely integrated paint mixtures. The presence of the pigments bone black and lead white (basic lead carbonate) on *Mt. Erebus* are, however, strongly indicative of the use of artist's, rather than household, paints.

Nolan's energy and drive for both travel and painting were often exhausting for those with him, particularly Cynthia Nolan during her struggle with tuberculosis in 1958–60. She wrote of him complaining to her, "Never tell me again that you're too tired to walk. Energy! Energy! Plug into the Almighty. I only need to go out and look up at the sky, see the ocean flowing or trees growing and I'm recharged. Can't you do that?" (C. Nolan 1967, 5).

The year 1964 was particularly productive for Nolan. He started the year exploring Antarctica by boat and helicopter and was then in Australia by February to attend the Adelaide Arts Festival, where his African paintings were being shown. By April he was back in London, and by September he had completed sixty Antarctica paintings and fifty paintings of Adelaide Ladies (Turley 2016). By December he was working on large

FIGURE 85
Sidney Nolan, *Peter Grimes*, 1977. Ripolin on board, 91.5 × 122.0 cm (36 × 48 in.). Private collection.

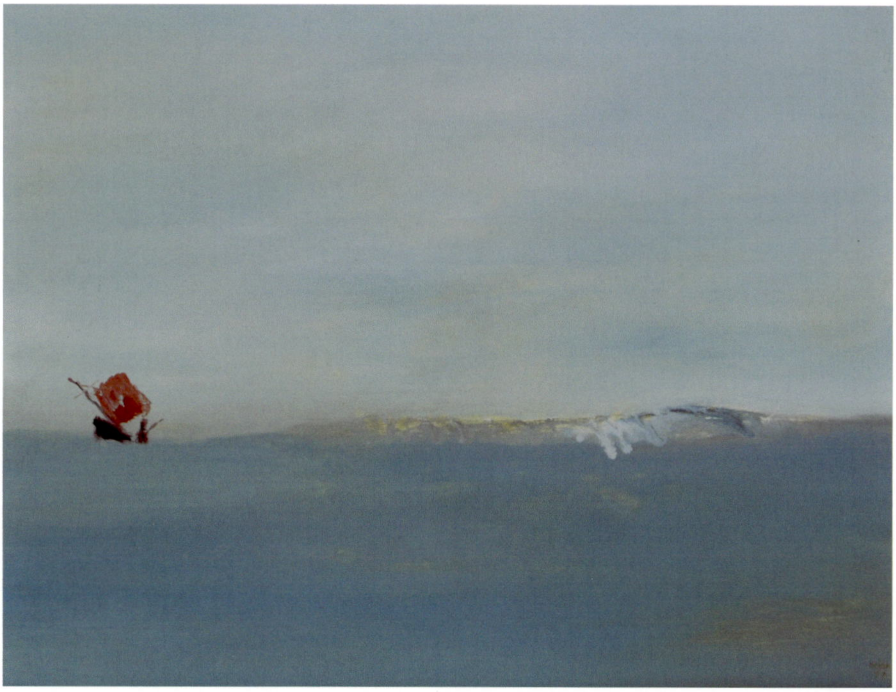

paintings of the Goulburn River, including the ten-meter multipaneled painting *Riverbend I* (1964–65), which he finished in January 1965.

Simon Pierse suggests that Nolan's prolific output was allied with automatic writing. The process of making dominates; paintings emerge with great speed and all are valued equally, as both successes and failures. While many other artists carefully edit their body of work by destroying their perceived failures, Nolan kept everything that he didn't sell or give away. "I put everything on the fact that I'll get it right, but sometimes I don't...if you don't get it right like that, then there is no second chance... I don't overwork much, just discard them and keep them. I'm just as fond of them as the others" (Nolan in conversation with John McDonald 1992; quoted in Klepac 2007, 83). According to Pierse (2017, unpaginated), this was "his way of putting down that instinctual first vision, in the full strength of its impact, before it could be alloyed with too much rational thought."

In 1976, Cynthia Nolan committed suicide. This unexpected and tragic event precipitated a period of inaction on Nolan's part. A rare painting from the period in his old medium Ripolin is *Peter Grimes* (1977; fig. 85). Fittingly, it depicts the moment in the opera by Benjamin Britten when Grimes scuttles his ship and drowns, overwhelmed by the hatred of locals who are enraged by his involvement in the accidental death of two young men (Lipsey 2017).

Shortly after Cynthia's death, Nolan left London and moved to live in Herefordshire at Mary Perceval's home, Ruthland. Nolan and Mary were married in 1978, and five years later they purchased the estate in Wales called The Rodd. This was Nolan's last home, but the decade spent there saw one more bout of prolific activity, driven by his take-up of a material, the spray can, that harked back to the beginning of his life and his work as a commercial artist.

Synthetic Emulsion Paints

First used in the house paint industry, synthetic resins emulsified with surfactants and other additives proved a technical watershed, particularly for do-it-yourself home renovators. Synthetic household paints based on at least four different resins were developed in the late 1940s and 1950s. Stryrene-butadiene and polybutadiene acrylonitrile were synthetic rubbers developed by the United States and Germany during the Second World War that found further use as paint binders in the immediate postwar years but were quickly superseded by PVA and acrylic paints. These early synthetic rubber binders are, however, the reason that emulsion paints became colloquially known as latex paints, even though most were soon based on PVA and acrylic (Standeven 2011).

PVA was the first synthetic emulsion binder used in house paint in Britain, available in the late 1940s but probably not used broadly until the mid-1950s (Standeven 2011). Acrylic emulsion paints, for their part, first became available in the United States in 1953, but they did not enjoy widespread use in Britain and Australia until the 1960s as they were initially prohibitively expensive.

The development of acrylic emulsion as an artist's paint in 1954 did not lag far behind its use in house paint. Although available as a binder for artists to make their own paints with from the 1950s, PVA was not developed as an artist's paint until 1966. By this time, artist's acrylic paints were dominant and there seems to be little recorded use of manufactured artist's PVA paints. Flashe®, a French-made artist's paint, is, however, a popular PVA paint still available to artists today.

The polymer types found in synthetic paints can be distinguished using FTIR spectroscopy on microsamples (fig. 86). Of the three different types of emulsion found on Nolan's paintings or in his studio, an externally plasticized PVA (fig. 86a) is distinguished in the fingerprint region (600–1500 cm^{-1}) by a broad strong absorbance at 1241 cm^{-1} and a doublet at 1435 and 1373 cm^{-1} in which the lower wavenumber peak is stronger in absorbance relative to the higher. The acrylic copolymer ethyl acrylate/methyl methacrylate (EA-MMA) identified in Nolan's tubes of Liquitex gel medium (fig. 86b) demonstrates a strong absorbance at about 1150–1170 cm^{-1} (with a doublet at 1176 and 1161 cm^{-1}) and at 1238 cm^{-1}. The doublet in the area of 1450 and 1380 cm^{-1} shows a reversal of relative absorbance compared to PVA. The FTIR absorbance spectrum of the polybutadiene acrylonitrile emulsion from the drum of Helizarin binder TS (fig. 86c) shows a small absorbance at 2237 cm^{-1} and two distinctive peaks at 969 and 804 cm^{-1}.

In addition to these polymer types, there were several emulsion paints developed in the 1960s in which PVA was internally plasticized with acrylic. FTIR analysis and identification of these is more difficult as they share features of both the PVA and acrylic FTIR spectra (Carter, Osmond, and Ormsby 2013). However, no examples of these types of PVA/acrylic copolymers were identified in Nolan's studio.

FIGURE 86

FTIR absorption spectra of three emulsion binders used by Nolan: (a) white PVA paint (externally plasticized) sampled from *The Galaxy* (1957–58; see fig. 74); (b) acrylic Liquitex gel medium (ethyl acrylate/methyl methacrylate copolymer) sampled from the Rodd studio; (c) nitrile rubber (polybutadiene acrylonitrile) Helizarin binder TS sampled from the Rodd studio.

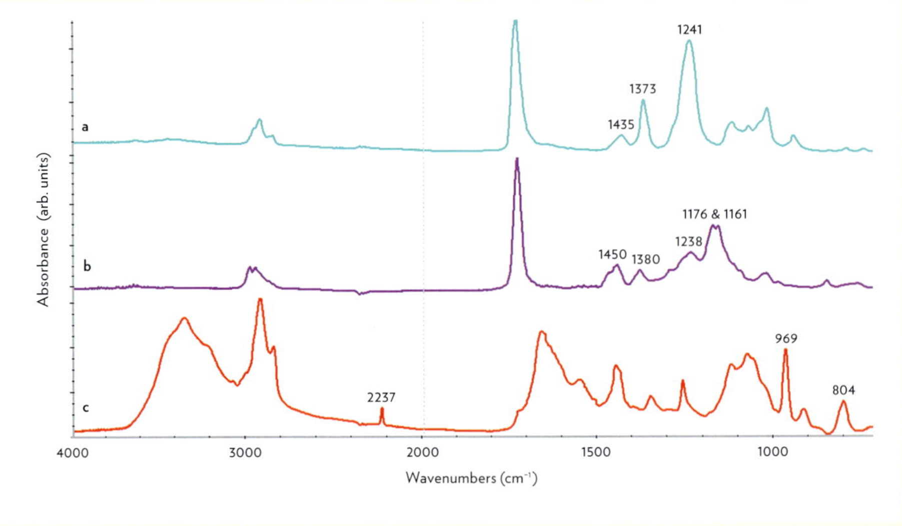

6 The Last Studio: 1983–1992

Nolan and his wife Mary purchased The Rodd, on the Welsh border with England, in 1983. The estate encompasses a seventeenth-century manor house and farmlands with a series of antique barns and stables. This last decade of Nolan's life was a productive one, not only for the paintings produced but also for the new materials and techniques he added to his repertoire.

Stepping into Nolan's last studio, in a partitioned section of a barn, is an immersing experience. The small space is dominated by a large flat table, and the shelves are full of painting materials, with spray cans arranged on two side tables for easy access to the range of colors (fig. 87). It is, typically for Nolan's work spaces, cramped and utilitarian.

This was not Nolan's only working space at The Rodd. His first studio was in an upper room of the house, the floor protected by bright blue plastic sheets. However, the toxic solvent of the spray paint soon drove him out to the adjacent barn. Nolan took well-advised precautions with the solvents, wearing protective gear, as a Polaroid "selfie" (fig. 88) and masks, goggles, and filters still on the shelves attest. Many of the paints now in the barn studio were originally in the house studio. Others came from

FIGURE 87
Alex Ramsay, Nolan's studio at The Rodd, 2017.

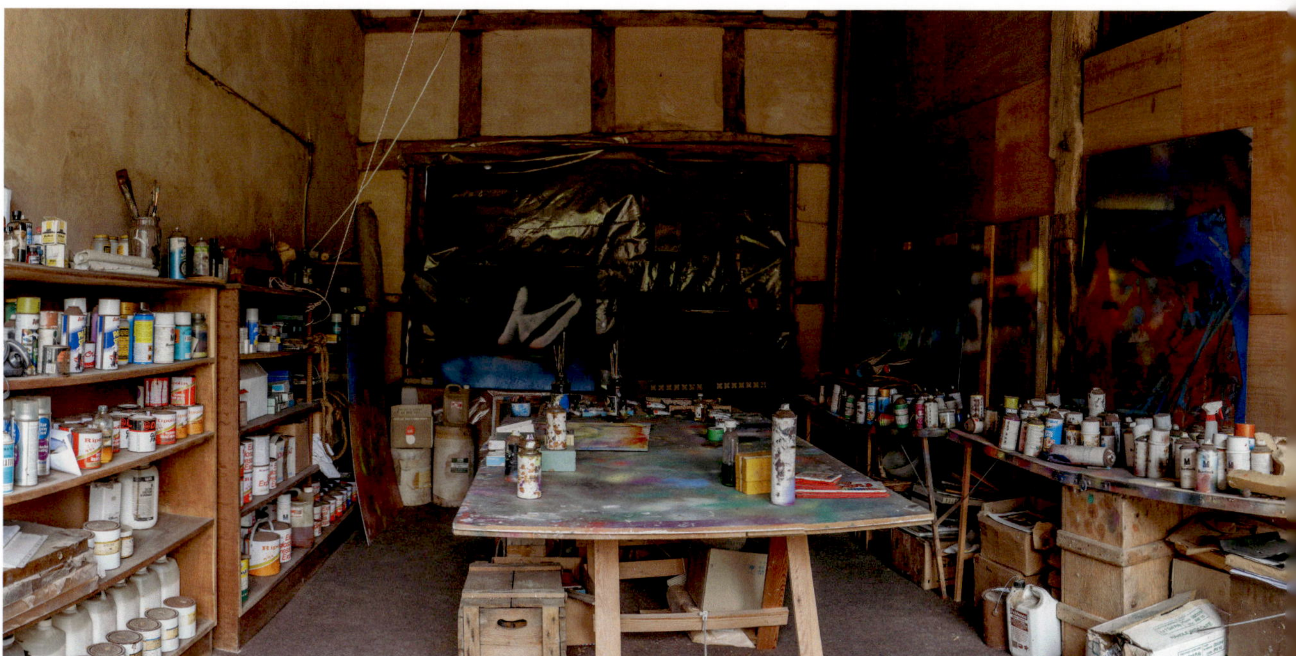

FIGURE 88
Sidney Nolan, Polaroid "selfie": Masked up for spray-painting, May 11, 1986.

FIGURE 89
Photographer unknown, Sidney Nolan outside the barn studio at The Rodd, 1986. The Sidney Nolan Trust.

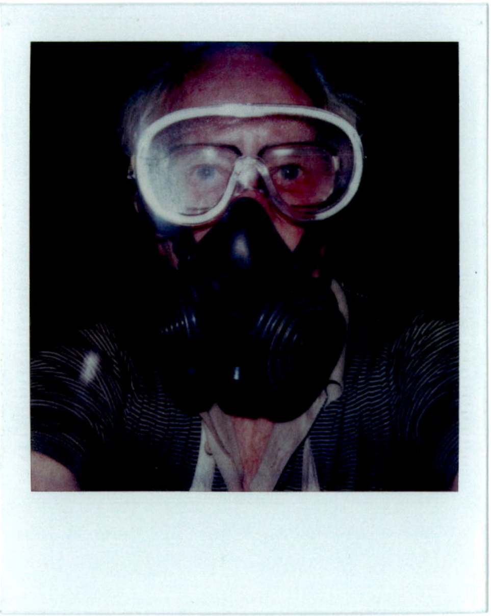

earlier studios in London, such as the Whitehall flat he used from the 1970s until the 1980s.[1] A larger barn across from the studio served as a work space for big projects as well as an exhibition gallery and performance space for concerts.

When Nolan first moved into the barn, he sealed the openings at both ends with carpets and plastic to keep out the wind and cold. This must have been only partially successful, for even today the wind flaps at the plastic still in place at one end. A photograph of Nolan outside the studio doors shows him well rugged up against the cold, wearing his usual butcher's apron for painting over his warm layers (fig. 89). Closing in the space must also have prevented him from looking out at the beautiful landscape of the Welsh Marches. Yet Nolan rarely worked directly from life, and this landscape

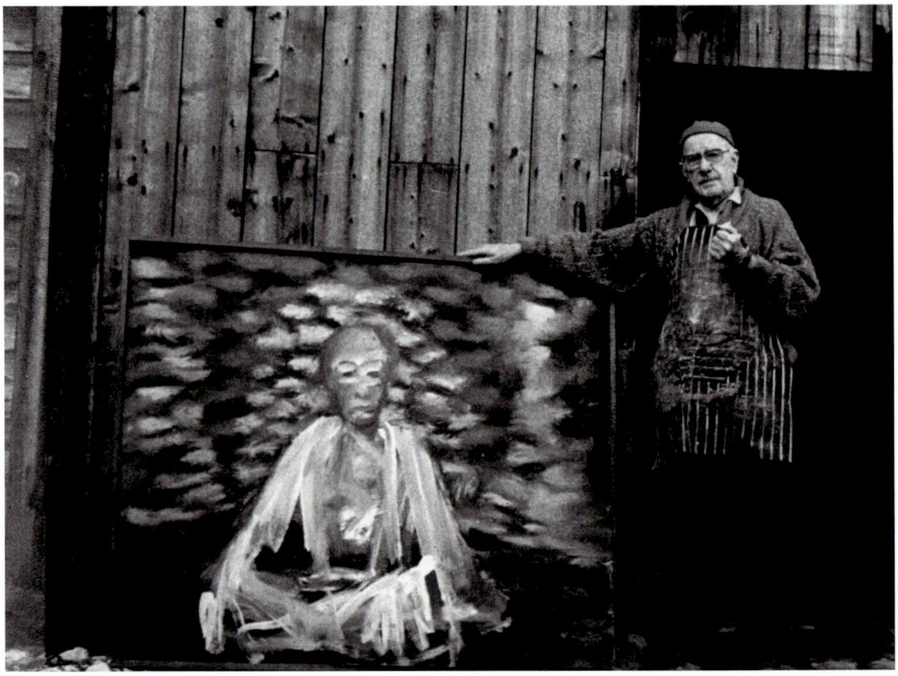

in which he lived for almost a decade never appeared in his paintings. Instead, Nolan worked under a bright mercury vapor light hung from the ceiling. This light emits a strange and cold green color quite unlike natural light, a strange choice for a painter. Although we know Nolan's preference for artificial rather than natural light throughout his life's practice, it is difficult to assess the effect of this mercury vapor light on his late paintings. Certainly, naturalism in color was never a focus of Nolan's work.

While many of Nolan's paintings dated prior to his move to the UK in 1953 have been analytically tested as described in the first chapters of this book, the following four decades have not yet been similarly studied. The studio therefore offers some important initial clues to his practice in Britain, though further work is required to relate these findings to the paintings. A full inventory of the more than one thousand individual items in this studio is not included here; however, appendix II, table 5, does list a selection of the spray paints, clear coatings, Helizarin dyes, Ripolin paints, and other paint media to be mixed with pigments that remain at the Rodd studio, along with an analysis of their contents.

Remnants of Australia

Under the main painting table at the front of the barn studio is a wooden box originally made for transporting a World War II machine gun from Britain to Australia. Tucked within the retained straw packing is a receipt for a consignment of Vickers guns sent to the Royal Australian Air Force (RAAF) in 1943 from Britain's Royal Air Force.[2] Whether Nolan "requisitioned" this box during his time in the Australian Army, when he was based near the RAAF airfields in the Wimmera, or purchased it after the war as surplus is unknown. Marked on the lid "Sidney Nolan, Artist's Materials" and "not wanted on voyage," it contains remnants of items sent from Australia in 1950, when Nolan and his wife Cynthia shipped their Hillman Minx motor car and traveled around Europe.[3] The items include spare bulbs and rotors for the car and a methylated spirit camping stove, presumably in anticipation of the long car trip they took through France, Spain, and Italy. One can of Ripolin paint, Canary Yellow gloss, identical in type to those left in the Wahroonga studio, is a solitary specimen of the paint he used from 1943 to 1953.

A collection of odd elliptical-shaped cans also found in the wooden gun case may hold clues to some of Nolan's very first paintings (fig. 90). FTIR analysis of the drips on the outsides of the cans labeled "Glaze" and "Stand" show that these contain drying oil, while the contents of the cans labeled "Red Flo" and "Black Flo" are based on ester gum. Ester gum is a modified form of rosin, also called colophony. This natural resin,

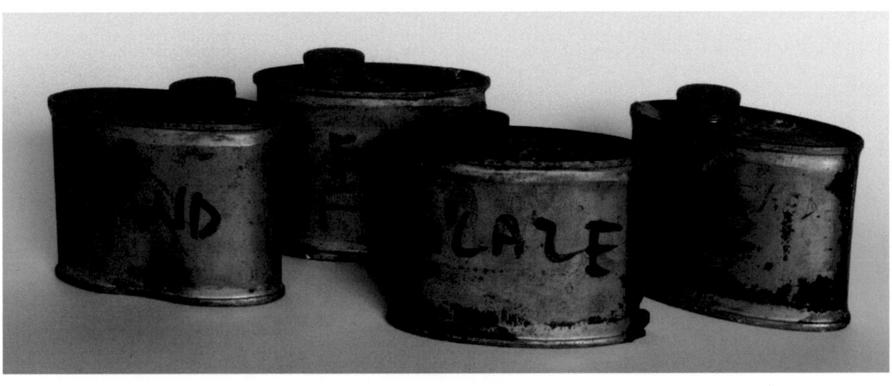

FIGURE 90
Group of elliptical cans of paints from gun case at Nolan's studio at The Rodd, 2017.

exuded from pine trees, is deeply colored and known for brittleness, but it was a popular paint additive in the 1920s and '30s when there were few alternatives for cheap gloss paints. Ester gum was often used with tung oil (Standeven 2011, 29) to make very fast-drying paints, and this may be the type of oil present in the can of "glaze."[4] Tung oil, an extract from the tung or China nut, was a versatile commercial paint binder that dried much faster than traditional linseed oil. When mixed with ester gum, it formed quick-drying glossy and transparent paints. Ester gums are also compatible with nitrocellulose, which Nolan was using by 1941. It has not been possible to confirm with analysis the use of these paints on a small group of abstract works from 1940, but the color and gloss transparency of the drips of "Red Flo" and "'Black Flo" on the cans show similarity to the red and black paint layers on *Abstract* (c. 1940; Heide Museum of Modern Art), for example. Perhaps Nolan, having made his first paintings with these ester gum paints, felt attached to them and packed them for his trip to the UK. Several other similar tins remained in his studio in Wahroonga, Sydney.

Ripolin

Ripolin continued to be an important medium for Nolan in the four decades he spent living and working in Britain. However, the Ripolin paint Nolan purchased in London was a variation on standard Ripolin, called Ripolin Express™. Made with alkyd resins, Ripolin Express was first manufactured by the French Ripolin company as early as 1936 (Gautier 2011, 128), but it appears that Nolan did not obtain this type of Ripolin paint until after his move to London.

The cans of Ripolin in the Rodd studio are found in two locations. One group, on the studio shelves, comprises thirty-five cans of Ripolin Express in nineteen colors, sixteen of which were sampled for analysis. The color names on these cans, for example Boy Blue, Cote d'Azur, and Goldfinch, in contrast to the more prosaic prewar designations such as Blue Deep 40RD and Red 16. The Ripolin Express paints represent a range of muted tones dominated by blues and greens, quite unlike the bright primary colors that Nolan had used in the 1940s. The new Ripolin selection may have been inspired by the color shift in his new country, even if the landscape itself did not appear often as a subject for painting.

The binder in the shelved Ripolin Express cans was identified with FTIR analysis as alkyd (see appendix II, table 5). The pigments identified in the paints also mark a change from the prewar Ripolin paints. No longer based on the white pigment zinc oxide, these Ripolin Express paints now utilize titanium white (titanium dioxide). The colored tints were generally obtained using synthetic organic colors, and chrome yellow (lead chromate) is no longer present, having been substituted by several organic coal-tar yellow colors.

A second group of Ripolin cans was found in a stack of cardboard boxes at the front of the studio, perhaps representing an earlier production date than the shelved Ripolin Express. These boxes contain a mixture of materials, including two prewar cans of Ripolin (gloss White 1 and Pink 12), seven cans of an earlier type of Ripolin Express than those on the shelves, and several cans of the later type. No sampling or analysis of this second batch of materials has been undertaken to confirm whether these earlier Ripolin Express paints are also alkyd-based. It would be useful to know if they, too, made use of the popular postwar white pigment titanium white. This is a project worthy of investigation, as identified shifts between the prewar Ripolin that Nolan was using in Australia and the cans of Ripolin Express in his studio at The Rodd may provide useful analytical markers to distinguish between paintings made pre- and post-1953. In addition, assigning medium descriptions to post-1953 Ripolin paintings like *Peter Grimes* (1977; see fig. 85) should consider the possibility that these may be alkyd paints, not oil enamels. Although the studio contains no flat Ripolin paint as Nolan had used in Australia, other Ripolin products on the studio shelves include polyurethane varnishes (both gloss and eggshell), vinyl matte emulsion white paint, and texture coating. Future analytical investigations of Nolan's paintings may find that these types of materials were used as priming layers and varnishes.

Spray Cans

In 1982, the year prior to moving to The Rodd, Nolan painted thirty-one works for a film about the poet Arthur Rimbaud, *It Is of Eden I Was Dreaming* (1983, directed by David Muir), using spray paint in aerosol cans. This set Nolan on a new technical pathway investigating the potential of these richly tinted acrylic paints. No doubt he took to the sprayed paint naturally, harking back as it did to his first jobs painting advertising sign-age, but now he wielded the cans two-handed like a double set of guns firing at the canvas. Although the paint had to be used in the premixed colors, a film of Nolan painting in 1987, *Sidney Nolan: Such Is Life* (directed by Brian Adams), shows his remarkable technique, the droplets of paint meeting in midair before falling onto the canvas with a wet merging. The spray cans produced spectacular effects of colored sky and water, likenesses in portraits, and huge abstracts created using torn cloth as stencils and pressed into the still-wet paint for textured effects (fig. 91). Boxes of these rags with colored overspray remain under the work table in the studio.

The spray cans dominate the barn studio, set out on two side tables for access to a range of colors (fig. 92). There are different brands and types of spray paint, but by far the most plentiful are the German paint Marabu in thirty colors, called Buntlack™. Twenty-five of the various spray can colors were still viable in 2017 and spray samples were taken for analysis (see appendix II, table 5). The paints are based on the fine and dense white pigment titanium dioxide mixed with a range of organic and inorganic pigments, including fluorescent dyes, offering a brilliant palette of intense bright colors with an extremely matte finish. The pigments are dispersed in an acrylic binder that was analytically identified as isobutyl methacrylate. The acrylic in the Buntlack spray cans is dissolved directly in a fast-evaporating aromatic hydrocarbon solvent. Solvent solution paints dry as soon as the solvent evaporates, shortly after the paint droplets are expelled from the can and hit the canvas. When the can is held at a distance from the painting support, the paint particles are partly dry before contact, giving incredibly matte but powdery surfaces.

One of the paintings from the 1983 film *It Is of Eden I Was Dreaming*, and therefore one of the first works executed in this new medium, is titled *I Shall Tattoo Myself* (1982; fig. 93). It is a large stretched canvas worked in a limited number of the Buntlack colors, which can be matched visually with color samples sprayed from the cans: Lemon Yellow 020, White 070, Vermilion Light Red 030, Black 073, Fluorescent Green 364, and Pink 030. Jane Clark (2017) has identified the title of the painting as deriving from a poem by Arthur Rimbaud, *A Season in Hell, Delirium I: The Foolish Virgin–The Infernal Spouse*, written in 1873, in which the subject angrily describes being enslaved in a violent and dysfunctional relationship. The spirals on the cheeks and the portrait from 1982 bear a remarkable likeness to the hidden portrait revealed by synchrotron-sourced XRF mapping on the Kelly painting of 1945 (see fig. 35 and discussion in chapter 3).

FIGURE 91
Alex Ramsay, Rags used as stencils and dabbed into wet spray paint in Nolan's studio at The Rodd, 2017.

FIGURE 92
Alex Ramsay, Table of spray cans and model boats used for America's Cup paintings in Nolan's studio at The Rodd, 2017.

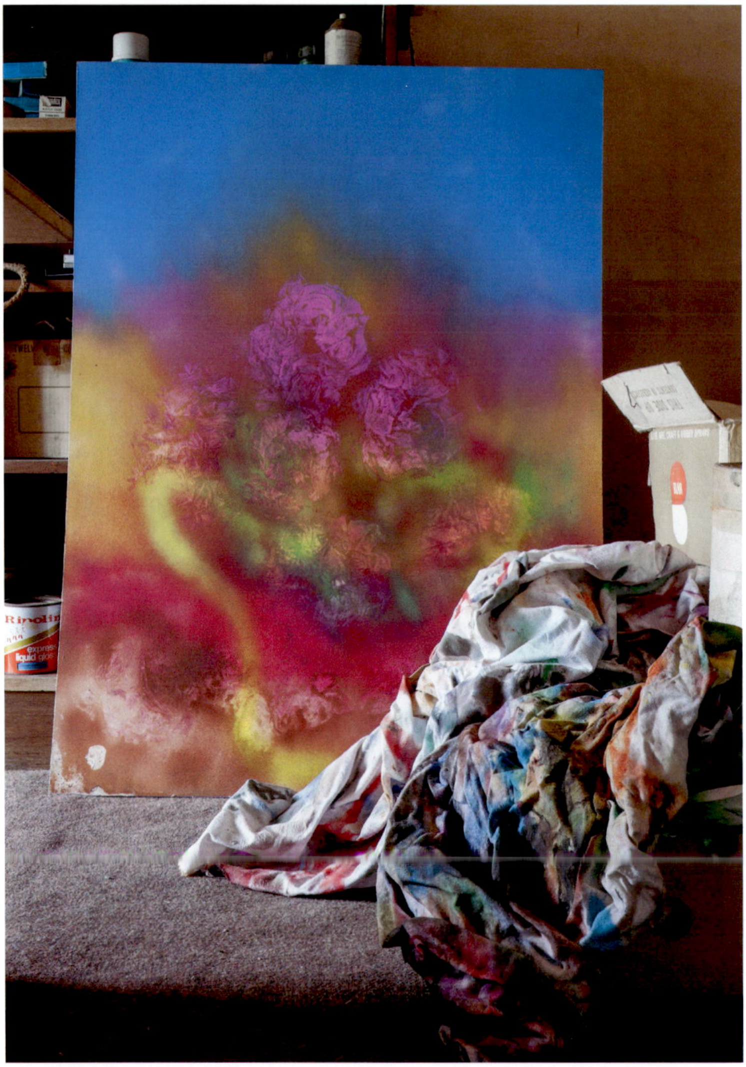

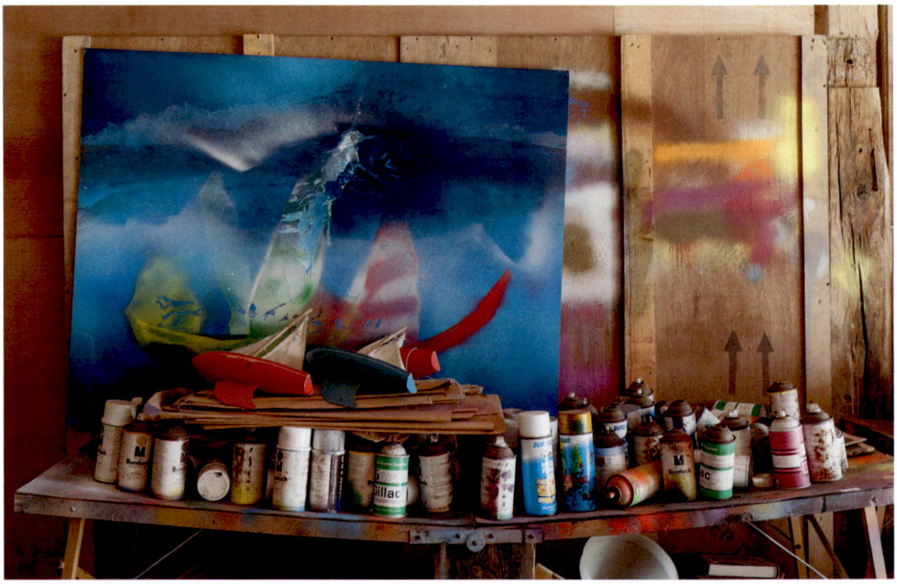

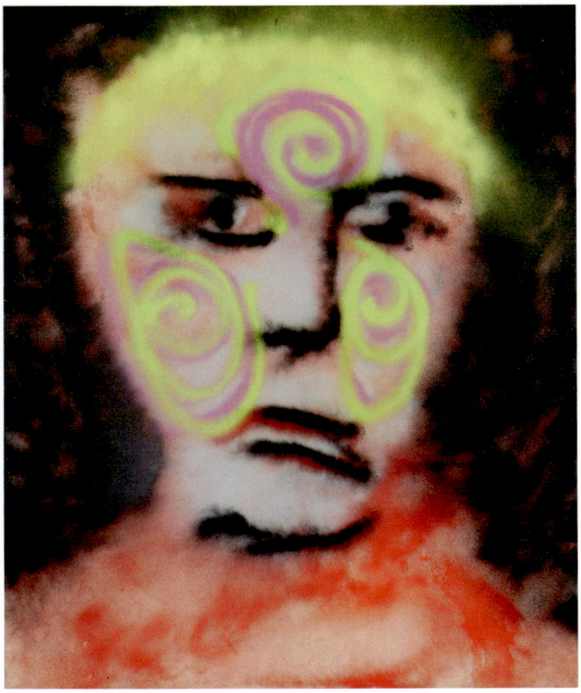

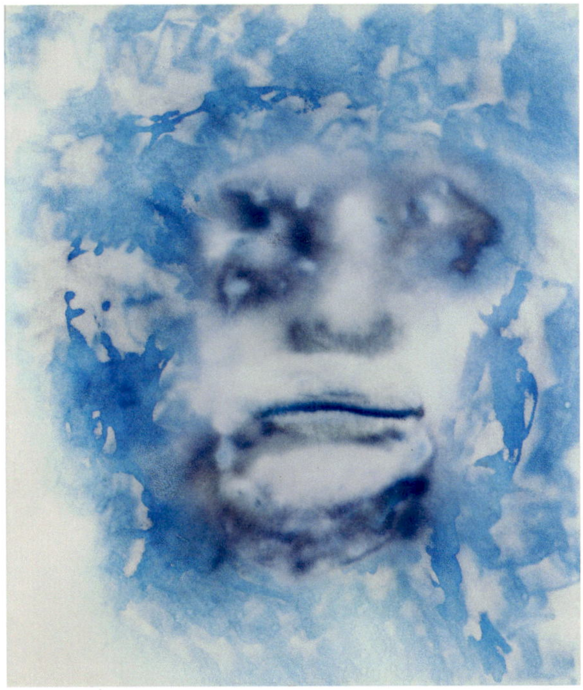

FIGURE 93
Sidney Nolan, *I Shall Tattoo Myself*, 1982. Acrylic spray paint on canvas, 184.0 × 162.0 cm (72½ × 63¾ in.). Collection of The Sidney Nolan Trust.

FIGURE 94
Sidney Nolan, *Portrait of Brett Whiteley*, 1982. Acrylic spray paint on canvas, 184.0 × 162.0 cm (72½ × 63¾ in.). Collection of The Sidney Nolan Trust.

Did Nolan remember the head he had obscured with Kelly's helmet almost forty years earlier? Is this later incarnation a clue to a possible subject (or multiple subjects) of that earlier painting, and Nolan's personal transference through his painting into the figures Ned Kelly and Arthur Rimbaud? Nolan himself said of the portraits in spray paint, "I thought that they would turn out hazy and lyrical, but they were traces left in my brain, kind of half memories, lines in my brain" (quoted in Nolan Gallery 1987).

The spray paintings from 1982 have an overall gloss that is not a feature of the Buntlack paints, suggesting that clear lacquers were applied as finishing layers. Abundant boxes of Blair Spray Clear Protective Coating indicate that this was the most favored of the numerous types of clear coatings in the studio. An acrylic like the Buntlack, it would have been compatible with the colored layers and should have good lightfastness, but future solvent-based removal of yellowed coating layers may be problematic owing to the similar composition of coatings and paint layers.

As Nolan developed his spray-painting techniques, paint effects become increasingly difficult to understand by visual examination alone. Studio contents suggest that other materials were sometimes used in combination with the Buntlack, including metallic spray paints made for touching up cars and a diverse range of clear spray fixatives and coatings. Many of the spray coatings in the studio have been analyzed and identified as the same acrylic resin as in the Buntlack colored paints (see appendix II, table 5), but some, such as Frisk Lac™ gloss (ethyl cellulose) or Rowney Perfix™ (polyvinyl acetate), might provide the effects seen in paintings from the mid-1980s, particularly the Chinese landscapes from 1986. The exact processes involved are not clear, and many of the Buntlack works, such as *Portrait of Brett Whiteley* (1982; fig. 94), have mysterious resist patterns, separations, and pooling of paint across the surface in a complex interplay between colored and clear layers. Elwyn Lynn (1987, unpaginated) wrote that in the early 1980s portraits Nolan "dispersed the paint by spraying it and so diluted it and floated it with fixative that the paint became as thin as mist or as opaque as thunder

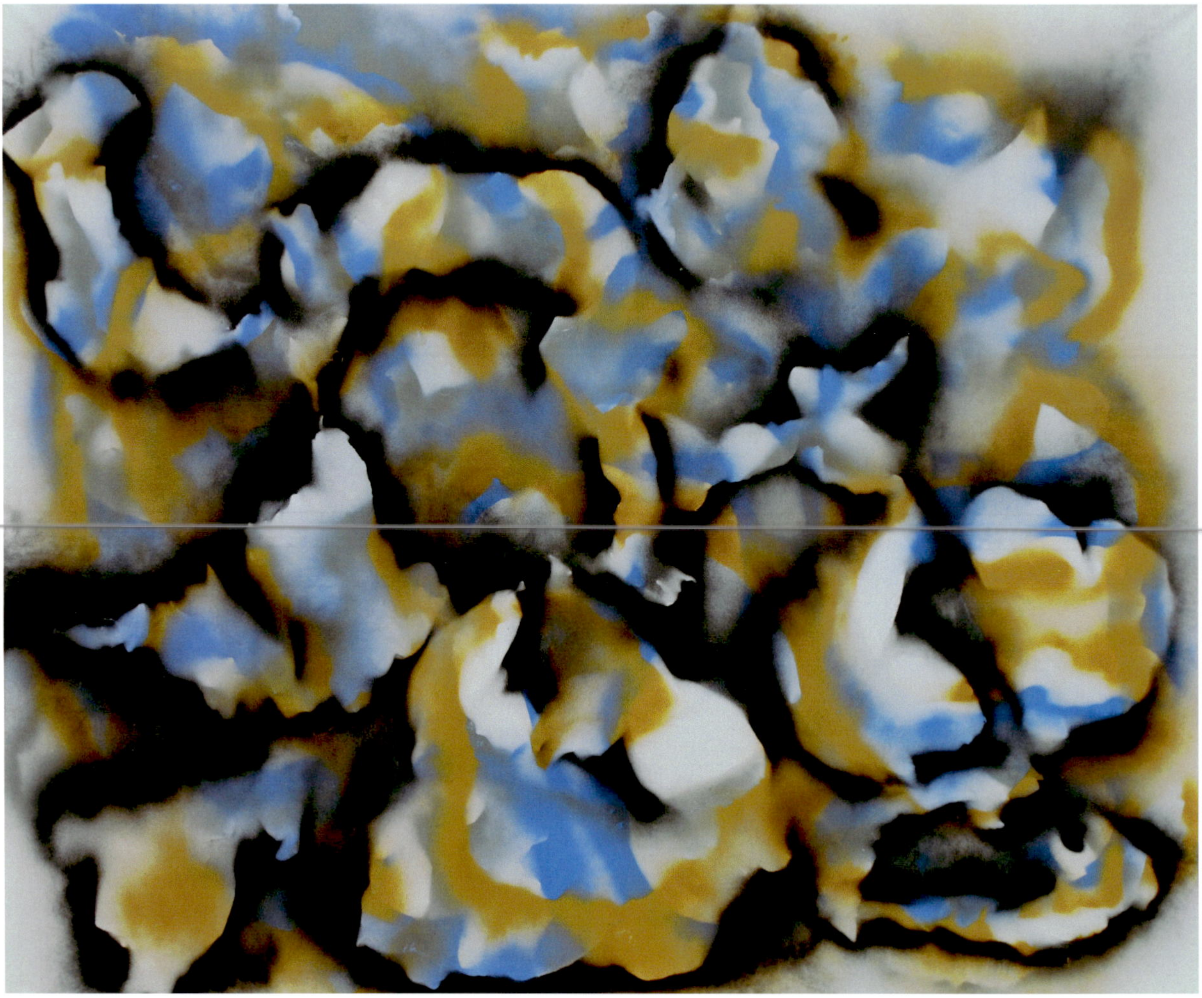

FIGURE 95
Sidney Nolan, *Abstract (Yellow)*, 1986. Acrylic spray paint on canvas, 198.0 × 245.0 cm (78 × 96½ in.). Collection of The Sidney Nolan Trust.

clouds." This suggests that Nolan may have been spraying clear coatings and colored paints together.

Whatever the exact process, it was messy. Overspray is everywhere in the studio—on tables, on paint cans, on walls, even on the plastic covers over the barn doors. As the spray-painting projects became bigger, Nolan moved into the gallery barn, even rigging himself a harness in which he hung from the ceiling, spraying while swinging over the canvas laid on the floor (Pearce 2007). Unfortunately, no photographs of this agile process have been found. Jane Clark (2017, 9) also recalled seeing him at work spraying paint as he lay across a wheeled trolley being pushed around by his studio assistant, John Hull. Now working on canvas up to four and a half meters (approximately 15 feet) long, he placed torn rags directly on the canvases and, suspended from above, sprayed around their edges. Working in at least three colors, each one with its own arrangement of rag stencil shapes, he created incredible abstracts that were strongly evocative of the Carnarvon Gorge rock art paintings he had seen three and a half decades earlier (fig. 95). He also stopped using clear coatings or working with resist techniques, leaving the surfaces of the large abstracts matte and finely textured only by the splatter of paint.

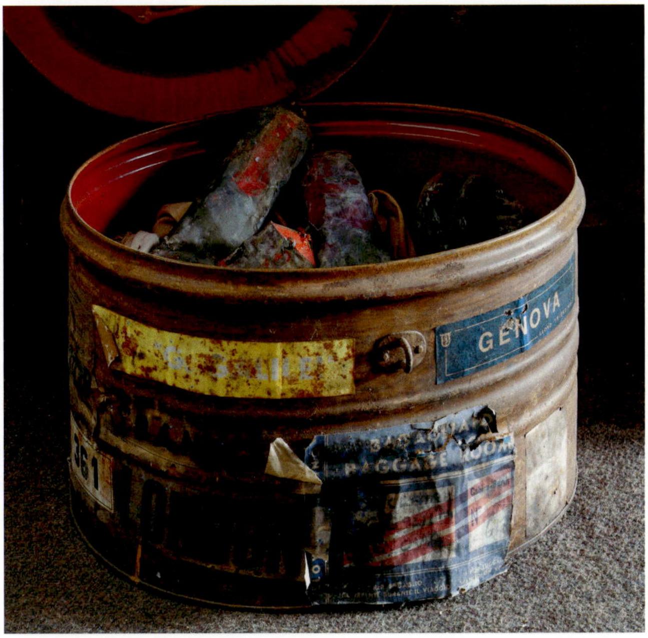

FIGURE 96
Alex Ramsay, Oval tin of oil paints made by Arthur Boyd in Nolan's studio at The Rodd, 2017.

A Return to Oil

A final surprise of the studio at The Rodd is the large number of oil tube paints, both laid out on the worktable on sheets of newspaper for access and stacked in boxes on the shelves. A metal oval box holds a collection of unlabeled tubes of colored paint, the distinctive finger-smeared swoosh of paint on the outside suggesting that they were hand ground by Arthur Boyd, Nolan's brother-in-law (fig. 96). In Australia, in the midst of wartime shortages, Nolan had mentioned to Sunday Reed that he might like to paint with oils in the future: "Still a hankering after oil-color but I really feel it belongs with Rembrandt & Cezanne—or after the war" (Nolan to Sunday Reed, c. October 1942).[5]

Nolan had used artist's oil paints in 1939, on the group of paintings on blotting paper, diluting them into thin slurries that the paper quickly absorbed. Later, in London in the 1960s, he mixed oils with alkyd gel medium to speed drying and make

FIGURE 97
Sidney Nolan, *Untitled*, 1989. Oil on canvas, 152.5 × 122.0 cm (60 × 48 in.). Collection of The Sidney Nolan Trust.

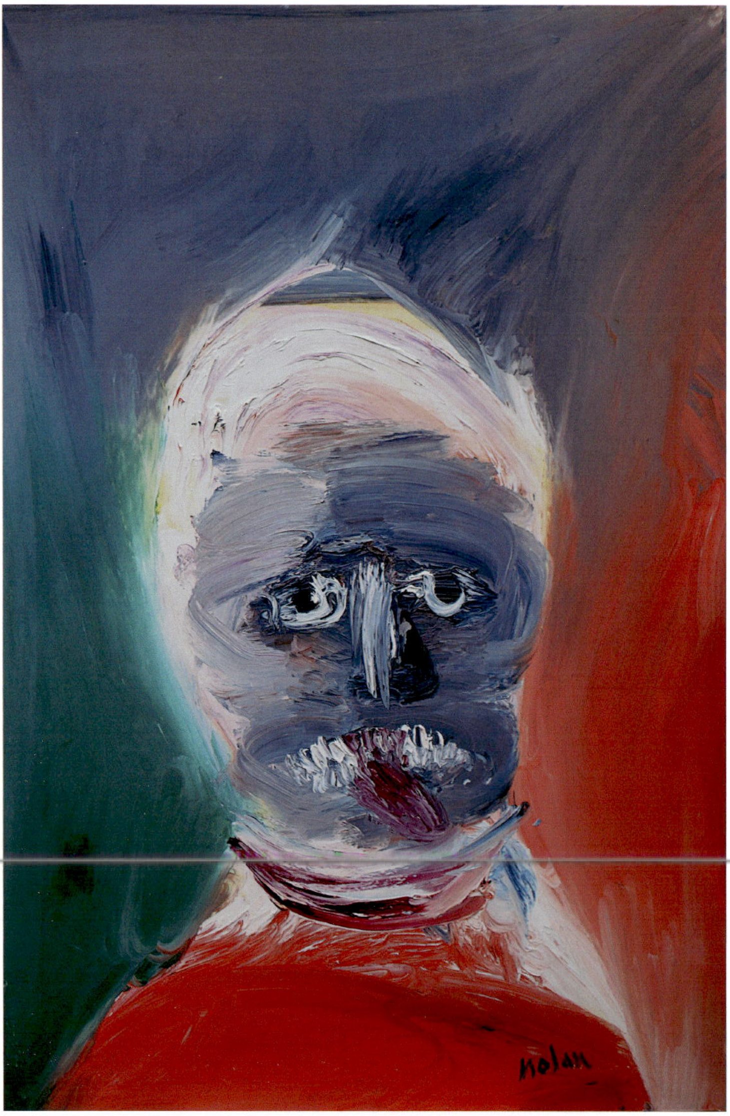

translucent paints. But rarely had he used them for their inherent thick, pastelike opacity. An exception was a group of heads from the 1980s (fig. 97). Brilliantly colored with backgrounds brushed in from all edges, as if Nolan was still working flat on a table rather than on an easel, these portraits of unknown subjects with tongues sticking out and crimson slashes across the neck suggest violent death.

Nolan's own end in 1992 was quiet and quick. Suffering a fatal heart attack after returning to his flat in London from a visit to the Tate Gallery, he lay back on the bed and was immediately gone, a life of incredible adventure in paint and travel completed.

To have two studios containing the materials of an artist's long working life is extraordinary. Why Nolan needed such abundance is perhaps hinted at by his wartime experience: "I have in the back of me the feeling we are going to need a lot of color before the new age" (Nolan to Sunday Reed, February 4, 1943). This wish is surely fulfilled by the storehouse he collected at The Rodd. Given the strong pull of his youth in his subjects for painting throughout his life, it seems fitting that this last studio contains paints from Nolan's earliest life as an artist, in the small elliptical cans containing ester gum enamels, as well as his last great material choice, pressure-pack spray cans.

7 Observations on Condition and Conservation

The most commonly asked question regarding the condition of Sidney Nolan's paintings is whether household paints are a stable artist's medium. Nolan's mentor Ralph Mayer didn't think they were when he reissued his *Artists' Handbook of Materials and Techniques* in 1970:

> The experimental painter for whose special requirements none of the traditional painting materials seem to be ideal, longs to explore this field, and before such products were made expressly for artists, he sometimes turned to industrial paints on the market, such as the Duco-type lacquers, the bright household enamels, and the mat finish interior wall paints—excellent products, but all foredoomed to failure because they are not intended for artists' use. (Mayer 1970, 198)

Whether Nolan would have taken heed even if Mayer's warning had been in the 1940 edition seems unlikely. In any case, Mayer's assessment may have been more dire than is suggested by this case study of Nolan's paintings, in which almost all available types and varieties of household and industrial paints have been identified.

The paint industry took the research and development of new materials very seriously. In addition to the do-it-yourself home decorator, they had to answer to well-trained and vocal professional painters' organizations. The vigorous debate and sharing of information is evidenced in industry journals such as *The Decorator & Painter for Australia & New Zealand,* one of the most comprehensive sources of information regarding commercial paints in the twentieth century. The work developing new materials was often undertaken by chemists, and products were rigorously tested both before release to the public and then on an ongoing basis, to ensure durability and performance.

In paint industry terms, a durable paint is one that is still functional after ten years of outdoor weathering—exposure to high levels of ultraviolet light, humidity, and heat. This is quite unlike the conditions experienced by most artworks, which are typically kept inside with reduced light levels and some degree of humidity control, and not generally in direct contact with water. For artistic use, therefore, household and other types of commercial paints should be expected to have excellent longevity and stability. This mostly appears to be true for Nolan's paintings up to seventy years after they were created. There are, however, a few important exceptions.

Perhaps the most vulnerable medium in Nolan's repertoire is the nitrocellulose lacquer that he began using in 1941. Cellulose nitrate is not considered to have good long-term stability, especially when exposed to light, humidity, or heat. Degradation is due to the cleavage of the nitrate group in the polymer and the formation of nitric acid, which can further degrade the polymer (Shashoua 2008). Breakdown of nitrocellulose polymer occurs internally with the formation of a network of microcracks that, over time, spread to the surface. The nitrocellulose may maintain its structural strength until it becomes so damaged that it disintegrates, thus disguising the real extent of the internal damage. Pigmented nitrocellulose paint appears to have greater stability than clear film, possibly due to the presence of white pigments that reflect, rather than absorb, the full spectrum of light. The addition of oil and alkyds in typical nitrocellulose lacquer formulations from the 1930s and '40s also assist in stabilizing the paint.

Nolan's paintings in nitrocellulose from 1941 generally seem to be in excellent condition, but this appearance may be deceptive. *Luna Park* (1941; see fig. 11) has a network of cracks through the body of the painting, indicating that the paint is fragile. This condition may have been caused in part by the cotton canvas support, a departure from Nolan's more usual solid supports for his nitrocellulose paintings, cardboard and slate. Woven cotton supports expand and contract with changes in humidity and are more vulnerable to impact damage. *Luna Park* has been lined onto an additional canvas, possibly in recognition of the fragility of the paint and its support, but the inevitable heat involved in this process may itself have contributed to the further deterioration of the nitrocellulose paint. This work was one of the most challenging for analysis, as the paint was so fragile it crumbled into powder when sampled, signaling that it has in fact developed extensive microcracking and degradation. Unfortunately, once this process has been initiated it continues, even when the catalysts of light, heat, and moisture have been removed. The rate at which this breakdown continues may, however, be inhibited. Typically for nitrocellulose, storage recommendations are for low light levels of less than 50 lux, cold temperatures (2–5°C), low relative humidity (20–30%), and exclusion of ultraviolet wavelengths of light (2006 Danish National Cultural Heritage Agency guidelines, cited by Shashoua 2008). These conditions are unlikely to be met during exhibition and display of paintings, though. It is not clear how effective the use of UV-exclusion glass or acrylic glazing is, or whether the application of a varnish layer may slow down the aging processes. Conversely, sealing nitrocellulose paintings within enclosures may allow the destructive buildup of nitric acid, a by-product of the degradation reactions in the polymer, and is therefore not recommended.

In comparison, the paint films on works by Nolan in alkyd dated from 1942 to 1947 are in exceptional condition. Microsamples taken from these works showed the paints to be surprisingly tough and flexible. Likewise, Nolan's paintings in oil-based Ripolin seem to have aged well. The painting *First-Class Marksman* (1946; see fig. 30), for one, is in immaculate condition. It retains its unvarnished state, and the brightness and gloss levels of the paints appear as strong as those recently painted out from the same cans from which it was painted. The rest of the works from the pivotal Kelly group originally exhibited together at Melbourne's Velasquez Gallery in 1947 do show signs of damage, possibly due to poor environmental conditions, harsh touring, and suboptimal storage

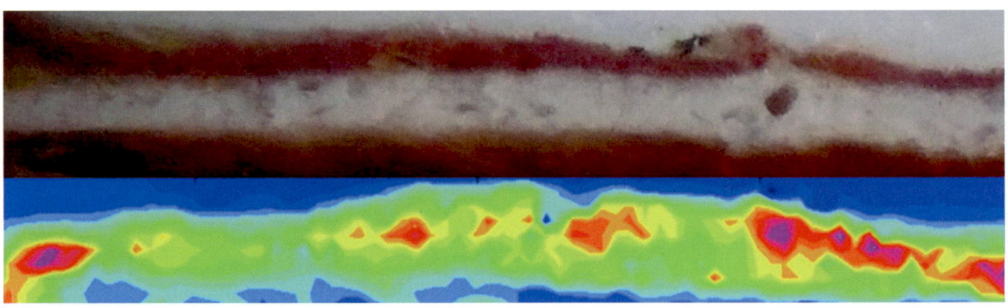

FIGURE 98
Micro paint sample from Sidney Nolan, *Burke and Wills Expedition, "Gray Sick,"* in incident light (top) and FTIR mapping (bottom) showing distribution in red of 1540 cm^{-1} absorption peak corresponding to zinc stearate/palmitate in white ground layer.

and conservation treatments, which included restoration and varnishing, before they were gifted to the National Gallery of Australia in 1977. The group of Kelly paintings from the same period in the Canberra Museum and Gallery are in better condition, perhaps because they have not had the same history of extensive exhibition, travel, and conservation treatment as the National Gallery group.

The combination of zinc white and oil, as found in the prewar Ripolin paints, is known to be a catalyst for one of the most intriguing changes in paintings, the formation of zinc soaps, and as such is the subject of extensive ongoing study (see sidebar on p. 71: Ripolin and Zinc Soaps). The fatty acids of the oil and zinc ions in the pigment usually combine quickly to form zinc carboxylates within the paint film. Over time, these species can crystallize and clump together, forming aggregates that may migrate toward the surface or to the lower levels of the paint film, either bursting through the surface of the paint film or causing separation between layers in a painting. Although these zinc soap aggregates were observed and characterized in the paint-outs from the Wahroonga studio cans of Ripolin, they were not analytically detected in Nolan's paintings themselves. This discrepancy may be due to the continued presence of unoxidized fatty acids in the liquid paint, whereas on the paintings these acids have oxidized with exposure to air. A recent study of a paint makers' notebooks from the 1940s suggests that the use of stand oil in zinc oxide–based oil enamels such as are identified in the Ripolin paints may inhibit the formation of crystalline zinc carboxylates (Carter et al. 2019). This function may be due to the transformation of the straight-chain fatty acids into cyclic fatty acids during the pre-polymerization of the oil.

Crystallized zinc carboxylates were detected in many of the ground layers on Nolan's paintings. Nolan's ground layers are often a complex mix of white pigments including lead white, zinc white, and titanium white (see appendix I) and are therefore not Ripolin paint but likely other types of household paints or his own mixtures. A micro paint sample taken from the foreground of *Burke and Wills Expedition, "Gray Sick"* (see fig. 73) was examined as a prepared cross section at the infrared beamline of the Australian Synchrotron, and the presence and location of the zinc stearate/palmitate absorbance peak at 1540 cm^{-1} was mapped across the sample (fig. 98). It does demonstrate that zinc soap aggregates have formed in the ground layer and although they had not migrated to either the top or the bottom of the layer they have potential to migrate in the future. The visual observations of seedlike protrusions in some of Nolan's Kelly paintings at the National Gallery of Australia is worthy of further investigation and may signal an inherent vulnerability to zinc soap aggregates that is yet to fully materialize in other works.

In Nolan's paintings completed up to his departure from Australia in 1953, the careful and deliberate pairing of flat and gloss paints, often on the same work, implies that they were not intended to be varnished. However, several jars of picture varnish were found in Nolan's Wahroonga studio, including a bottle of Winsor & Newton's Winton varnish, an early polycyclohexanone resin. A similar varnish was detected in the sky of *Central Australia* (1950; see fig. 69), indicating that Nolan may have deliberately and selectively emphasized the glossiness of the blue Ripolin by local application of varnish. Boot polish on the early painting *Head of Rimbaud* (1938–39; see fig. 7) also points to manipulation of the paint qualities to achieve specific levels of gloss and matte finishes with the use of wax. Picture varnishes and wax polishes are present in Nolan's studio at The Rodd as well, suggesting that his interest in these media continued after his move to the UK.

The presence of clear coatings is important in the assessment of a painting's condition, as these layers tend to be especially vulnerable to aging and deterioration. Typically, with traditional oil paintings, varnishes are expected to deteriorate over time and are commonly removed and replaced. When the varnish is intermingled with pigmented paint layers, however, or acrylic clear coatings are applied on top of acrylic paint, removal is no longer an option. Thus, the clear coatings in Nolan's late spray-paint works may yellow over time but are likely irreversible. Yet the dominance of acrylic resins in most of the spray coatings analyzed in the studio is heartening, for acrylic is considered to have the best long-term stability of any clear coating systems. Polyurethane, present in the studio as both a spray and brush coatings, is less reassuring in this respect.

Although PVA has a tendency to turn slightly yellow over time, the yellowing, as with nitrocellulose paints, is less obvious when the resin has been pigmented. Nolan's PVA paintings appear to be in excellent condition, though some have been varnished over subsequent years, imparting a higher level of gloss and diminishing the beautiful satin surfaces.

Nolan's use of recycled supports was an inventive approach to the lack of materials during the war, and it gave him pleasure to manipulate the surfaces with their interesting textures and previously printed or painted images. Cardboard and Masonite are remarkably stable and enduring. Their vulnerability is usually related to edges and corners being damaged when unframed, or bowing due to absorption of moisture at the back of the board. The pale-colored, unprimed Masonite that Nolan used in the late 1940s and early 1950s can show a halo effect at the edges of the paint, which may be waxes and oils, either in the board itself or in the Ripolin paint, staining and in some cases whitening the surface. The exact cause of this effect is not yet clear.

At a preexhibition examination of Nolan's spray-painted portraits from the 1980s at IKON Gallery in Birmingham in 2017, it was observed that many of the stretched canvases had distortions (draws) at the corners related to the absence of keys in the stretchers to tension the canvas at the corners. No stretcher keys were in place, and they had probably never been inserted, as a box of unused stretcher keys was found in the Rodd studio. This seems entirely consistent with Nolan's disinterest with most of the conventions of traditional painting practice.

Nolan's eclectic use of materials provides a rich opportunity for gaining insight into the developments in paint technology in the twentieth century and the range of possible variations in artistic practice in their application. A study of the paint materials left in his studios also suggests a need for caution when assigning medium descriptions to his works. Much analytical work remains to be done, both to thoroughly catalogue his practice and to allow effective and safe conservation treatment of his paintings. It is critical that the solubilities and vulnerabilities of his paint materials are fully understood before solvents, heat, and moisture are introduced to unidentified paints. FTIR spectroscopy is a well-established technique for characterizing organic materials, particularly synthetic paint binders. While it may not always provide clear results with artist's paints, it has proved successful on many samples from Nolan's paintings due to the predominance of commercial paints with a high proportion of binder. Care must be taken, however, when assigning binder types from microsamples studied with FTIR to a painting with layers of different materials or heterogeneous mixing.

Artists' studios are invaluable aids in unraveling the complex range of materials and paint technologies available to the modern artist. Yet ironically, the paints themselves are often the least valued elements of preserved studio spaces. To be sure, scientific analysis helps to identify the chemical structures of pigments and binders in paintings themselves, but important information can also be obtained from the paint in its wet and unoxidized form. The physical properties of studio paint stocks, as well as container labels, provide handling and application information that analysis alone cannot. While an increasing number of artists' studios have been protected and preserved, very few still contain the paint materials that were in daily use. It is hoped that this research on Nolan's materials will help to encourage future preservation, inventories, and studies of the paint contents in artists' studios as well as of the studios themselves.

Appendix I

Analysis of Nolan's Paints in Works from 1939 to 1964, with Methods of Analysis

TABLE 2 Nolan's Paints in Works from 1939 to 1964

\# FTIR
* XRF
^ microscopy
□ SEM/EDS
◊ py/GC-MS

AGNSW Art Gallery of New South Wales, Sydney
HMMA Heide Museum of Modern Art, Bulleen, Victoria
QAGOMA Queensland Art Gallery | Gallery of Modern Art
UQAM University of Queensland Art Museum, Brisbane

Works not cross-referenced as figures are not reproduced in this volume.

Painting	Sample site	Pigments	Binding media
Untitled (Abstract) c. 1939 AGNSW 170.2013	White ground	Chalk #	No result #
Untitled (Abstract) 1939 AGNSW 172.2013 Fig. 4	Recto, black	Bone black, chalk #	No result #
	Recto, red	Chalk, dinitroanaline red (PO5)	No result #
	Recto, blue	Ultramarine #	No result #
(Untitled-Abstract) c.1939 AGNSW 132.2006 Fig. 9	Recto, lower center, unpainted photograph	Silver, calcium, iron*	No result #
	Recto, lower edge, center, blue	Ultramarine blue, gypsum, zinc white[1] (zinc carboxylates with stearate/palmitate) #^	No result #
	Recto, left edge, center, bright red	Chalk, dinitroanaline red (PO5) #*	No result #
	Recto, lower right, yellow	Chrome yellow, gypsum #*	No result #
	Recto, lower right, black	Bone black #*	No result #
Head of Rimbaud 1938–39 HMOMA 1980.66 Fig. 7	Verso, dark yellow	Yellow ochre, kaolin #	No result #
	Verso, pale yellow	Chrome yellow, gypsum #	No result #
	Verso, brown	Barites, kaolin, gypsum #	No result #
	Verso, blue	Ultramarine #	No result #
Head of a Woman (Luna Park) Sept. 1940 AGNSW 166.2013	Recto, blue	Prussian blue #	Oil (resin) #
	Recto, upper left, red	Kaolin, Prussian blue, (red not identified) #	Oil (resin) #
Luna Park 1941 AGNSW 35.2003 Fig. 11	Recto, Black	Bone black #	Nitrocellulose #
	Recto, Blue	Prussian blue, titanium dioxide #	Nitrocellulose #
	Recto, Yellow	Chrome yellow, titanium white*	Not analyzed
	Recto, Red	Titanium dioxide, (red not identified)*	Not analyzed
	Recto, White ground	Titanium dioxide #*	Nitrocellulose #
Bird Dec. 4, 1941 HMOMA 1982.231	Recto, blue	Prussian blue #	Nitrocellulose #
	Recto, red	Titanium white, barites #	Nitrocellulose #
	Recto, black	No result #	Nitrocellulose #

Painting	Sample site	Pigments	Binding media
Untitled (Figure and Bird) verso Dec. 11, 1941 UQAM 1977.23 Fig. 10	Verso, pink fragment in packing	No result#	Nitrocellulose#
	Verso, orange fragment in packing	No result#	Nitrocellulose#
Untitled (Round Tree) 1941 AGNSW 167.2013 Fig. 12	Recto, black	No result#	Nitrocellulose#
	Recto, pale blue	Prussian blue, titanium white, talc#	Nitrocellulose#
	Recto, red	Titanium white, talc, alizarin (NR9)#	Nitrocellulose#
	Recto, white	Titanium white#	Nitrocellulose#
Landscape Aug. 29, 1942 AGNSW 180.2013	Verso, dark blue paint	Prussian blue, gypsum#	Alkyd#
	Recto, deep red left edge	Chlorinated para red (PR4), barites#	No result#
	Verso, pale blue	Titanium white, (blue not identified)#	Alkyd#
On the Beach Oct. 28, 1942 AGNSW 182.2013 Fig. 19	Recto, white tent upper right	Titanium white#	Alkyd#
	Verso, blue gloss lower edge	Prussian blue#	Alkyd#
	Verso, red center	Chlorinated para red (PR4), barites#	No result#
	Verso, green center left edge	Prussian blue, chrome yellow#	Alkyd#
	Recto, black mixed with red upper center	Prussian blue, (black not identified)#	Oil, possibly mixed with resin#
	Verso, pale blue top edge	Prussian blue, titanium white#	Alkyd#
	Recto, yellow gloss center	No result#	Alkyd#
Nostalgia for the Sky Nov. 1, 1942 AGNSW 183.2013 Fig. 21	Recto, pale blue lower edge right	Prussian blue, titanium white#	Alkyd#
	Verso, pale green yellow, center	No result#	Alkyd#
	Recto, upper left, black	Chrome yellow, (black not identified)#	Oil, possibly mixed with resin#
	Recto, upper right, green	Titanium white, (green not identified)#	Alkyd#
	Recto, lower edge center, matte white	Titanium white#	Alkyd#
Farmer, Dimboola Nov. 2, 1942 AGNSW 184.2013 Fig. 20	Recto, right edge center, pale blue	Prussian blue, titanium white#	Alkyd#
	Verso, lower edge, brown	Chlorinated para red (PR4)#	Alkyd#
	Recto, lower left, yellow	Chrome yellow, lead sulphate#	Alkyd#
	Recto, center, matte white	Titanium white#	Alkyd#
	Recto, upper left edge, orange	Chrome yellow#	Alkyd#
Dimboola Dec. 5, 1942 HMOMA 1982.232 Fig. 22	Recto, upper right corner, pale blue	Prussian blue, titanium white#	Alkyd#
	Recto, center left edge, black	No result#	Oil#
	Recto, center right edge, white	Barites, titanium white, kaolin#	Alkyd#
Water Wheel, Luna Park 1942 HMOMA 1992.112	Recto, blue	No result#	Alkyd#
	Recto, yellow	Chrome yellow, lead white#	Alkyd#
	Recto, white	Lead white, chalk, barites#	No result#
Wimmera Landscape 1943 HMOMA 1980.70 Fig. 15	Recto, white ground	Lead white, chalk, kaolin, barites, zinc white (zinc carboxylates)#	No result#
	Recto, lower right corner, yellow	Chrome yellow/sulphate, zinc white (zinc carboxylates)#	Oil#
	Recto, upper left edge, pale blue	Prussian blue, zinc white (zinc carboxylates)#	Oil#
Self Portrait Mar. 1943 AGNSW 412.1997 Fig. 25	Side, center left, white ground	Chalk, barites, kaolin, lead white, zinc white (zinc carboxylates with zinc stearate/palmitate)#*	No result#
	Recto, black	No result (black not identified)#	Oil#
	Recto, upper left, red	Barites, toluidine red (PR3)#*	No result#
	Recto, lower center, gloss white	Zinc white*	Not analyzed
	Recto, lower right, gloss yellow	Chrome yellow*	Not analyzed

Painting	Sample site	Pigments	Binding media
Bathers	Recto, lower left edge, ground	Lead white, chalk, barites, zinc white (zinc carboxylates)#	No result#
Apr. 12, 1943			
HMOMA 1982.247	Recto, upper right edge, dark blue	Prussian blue, zinc white (zinc carboxylates)#	Oil#
Fig. 24	Recto, lower edge left of center, black	No result (black not identified)#	Oil#
	Recto, lower left edge, red	Toluidine red (PR3), barites, Prussian blue, zinc white (zinc carboxylates with zinc stearate/palmitate)#	No result#
Untitled (St. Kilda)	Recto, white ground	Chalk, kaolin, barites gypsum#	No result#
1943	Recto, blue sea	Prussian blue#*	Oil#
AGNSW 168.2013	Recto, pink sky	Zinc white (zinc carboxylates)#*	Oil#
	Recto, red boat	Toluidine red (PR3), barites#*	No result#
	Recto, green boat	Prussian blue, barites, arylamide yellow (PY1)	No result#
Boy in Township	Recto, black	(Black not identified), [zinc carboxylates]#	Oil#
1943	Recto, pink	Titanium white, carminic acid (NR4)#	No result#
AGNSW OA24.1962			
Fig. 26			
Ned Kelly: "Nobody knows anything about my case but myself"	Recto, white	Zinc white*	Not analyzed
1945	Recto, blue	Prussian blue, zinc white*	Not analyzed
HMOMA 1998.34	Recto, yellow	Chrome yellow*	Not analyzed
	Recto, red	Barites*	Not analyzed
Fig. 32	Recto, black	Cobalt (minor as drier)*	Not analyzed
The Camp	Recto, white ground, matte	Barites, talc, zinc white (zinc carboxylates with zinc stearate/palmitate)*#□	No result#
1946			
AGNSW 207.1978	Recto, lower left, bright blue, gloss	Prussian blue, zinc white (zinc carboxylates)*#	Oil#
Fig. 31	Recto, lower left, bright yellow, matte	Zinc white, chrome yellow, silicates*	Not analyzed
	Recto, lower left, red tree, gloss	Barites (red not identified)*	Not analyzed
	Recto, lower right, white gloss	Zinc white (zinc carboxylates)#*	Oil#
	Recto, lower edge center, upper black	No result (black not identified)#*^□	Alkyd#◊
	Recto, lower edge center, lower black	No result (black not identified)#*^□	Oil#
First-Class Marksman	Recto, lower ground, white	Barites, talc, zinc white (zinc carboxylates with stearate/palmitate)#*□	No result#
Dec. 12, 1946			
AGNSW 62.2010	Recto, upper ground, white	Zinc white, [chrome yellow]□	Not analyzed
Fig. 30	Recto, gloss white	Zinc white (zinc carboxylates)#*	Oil#
	Recto, blue sky, gloss	Prussian blue, zinc white (zinc carboxylates)#*□	Oil#
	Recto, blue mountain, gloss	Barites, copper phthalocyanine, zinc white (zinc carboxylates)#*	No result#
	Recto, red gloss	Barites, toluidine red (PR3)#	No result#
	Recto, black gloss, upper	No result (black not identified)#□◊	Alkyd#◊
	Recto, black gloss, lower	No result (black not identified)#□◊	Oil#
Colonial Head	Recto, center left edge, white ground	Barites, talc#^□	No result#
Apr. 4, 1947	Recto, upper right corner, gray ground	Chalk, barites, talc#^□	No result#
AGNSW 7.2001	Recto lower right edge, pale blue	Prussian blue, zinc white (zinc carboxylates)#^□	Oil#
	Recto, upper right, red ear	Barites, zinc white (zinc carboxylates), toluidine red (PR3)#	No result#
	Recto, lower center, dark blue stripe in tie	Prussian blue, zinc white (zinc carboxylates), barites#	No result#
Mrs. Fraser (originally titled *Woongoolbver Creek*)	Verso, white	Chalk, gypsum, kaolin	No result#
1947	Recto, lower dark green	Prussian blue, chrome yellow, barites [zinc stearate/palmitate]#^	No result#
QAGOMA 1995.134	Recto, upper edge left, lower white ground	Kaolin, talc, barites#^	No result#
Fig. 53	Recto, left edge center, upper white priming	Kaolin, talc, barites, zinc white (zinc carboxylates)#^	No result#
	Recto, center right, upper sky pale blue	Prussian blue, zinc white (zinc carboxylates)#^	Oil#
	Recto, upper right, lower dark blue sky	Prussian blue, zinc white (zinc carboxylates)#^	Oil#
	Recto, left of center, black hair	Prussian blue, chrome yellow, [zinc carboxylates], (black not identified)#^	Oil#
	Recto, lower edge right of center, gray tondo	Talc, kaolin, barites#^	Alkyd#
	Recto, upper right tree trunk, yellow pigment	Chrome yellow/sulphate#^	Oil#

Painting	Sample site	Pigments	Binding media
Island 1947 AGNSW 8.2001	Verso white	Chalk, gypsum, kaolin#*^	No result#
	Recto, lower dark green from side	Prussian blue, chrome yellow, barites#	No result (double carbonyl 1733, 1717)#
Fig. 51	Recto, lower edge center white ground	Kaolin, talc, barites, zinc stearate/palmitate#	No result#
Platypus Bay, Fraser Island Oct. 1947 QAGOMA 2013.154	Verso, white	Chalk, gypsum, kaolin#	No result#
	Recto, lower dark green	Prussian blue, chrome yellow, barites [zinc stearate/ palmitate]#	No result (double carbonyl 1733, 1717)#
	Recto, white ground	Kaolin, talc, barites, zinc stearate/palmitate#	No result#
Lake Wabby Oct. 1947 HMOMA 1982.258	Recto, lower green (khaki)	Iron-based red, Prussian blue, kaolin, chalk, zinc white (zinc carboxylates with zinc stearate/ palmitate)#	No result#
Fig. 52	Recto white ground	Kaolin, talc, barites, zinc white (zinc carboxylates)#	No result#
Indian Head (Remnant) Oct. 1947 Private collection	Verso, white	Chalk, gypsum, kaolin#	No result#
	Recto, lower dark green	Prussian blue, barites, zinc white (zinc carboxylates with zinc stearate/palmitate)#	No result (double carbonyl 1733, 1717)#
Fig. 56	Recto, white ground	Kaolin, talc, barites, zinc white (zinc carboxylates with stearate/palmitate)#	No result#
Burke and Wills, "Gray Sick" Nov. 30, 1949 AGNSW 445.1995	Recto, white ground	Titanium white, barites, chalk, silicates, zinc white (zinc carboxylates with zinc stearate/ palmitate)#*□	Oil#
Fig. 73	Recto, upper left corner, pale blue sky	Prussian blue, zinc white (zinc carboxylates with zinc stearate/palmitate)#	Oil#
	Recto, lower left edge, dark red foreground	Iron oxide#*	Oil#
Animal in Swamp 1958 AGNSW 93.2013	Recto, blue hoof	Prussian blue#	Polyvinyl acetate (externally plasticized)#
The Galaxy 1957–58 AGNSW 57.1974	Recto, dark blue, right edge of right panel	Prussian blue#	Polyvinyl acetate (externally plasticized)#
Fig. 74	Recto, thick white dots, right edge of right panel	No result#	Polyvinyl acetate (externally plasticized)#
Burke 1962 AGNSW OA1.1968	Recto, white ground	Titanium white, barites#	Polyvinyl acetate (externally plasticized)#
	Recto, brown, upper right	No result#	Polyvinyl acetate (externally plasticized)#
	Recto, off white, sky center right	Titanium white, barites#	Polyvinyl acetate (externally plasticized)#
	Recto, blue, sky upper right	Prussian blue#	Polyvinyl acetate (externally plasticized)#
Elephant 1963 AGNSW 438.1995	Recto, ground, foreground left	Titanium white, barites#	Polyvinyl acetate (externally plasticized)#
	Recto, red, center right	Barites, Prussian blue#	No result#
	Recto, green, center	Prussian blue#	No result#
	Recto, yellow, center	Chrome yellow, lead sulphate, Prussian blue#	No result#
Mt. Erebus Aug. 28, 1964 AGNSW 1.2011	Recto, gray, lower center	Bone black, lead white, zinc white (zinc carboxylates)#	No result#
	Recto, white, lower center	Lead white, zinc white (zinc carboxylates)#	No result#

1. Zinc white does not have a characteristic FTIR absorbance in the region of 4000–600 cm^{-1}. The beginning of a broad peak absorbance may be seen at 600–550 cm^{-1}, but its presence can also be inferred by the broad absorbance at 1500–1600 cm^{-1}, which indicates the formation of zinc carboxylate species in the presence of fatty acids from the oil and zinc ions from zinc oxide. Sharp absorbance bands at 1540 cm^{-1} indicates the further formation or presence of crystalline zinc carboxylate species such as zinc stearate and zinc palmitate. In this list, only pigments analyzed by FTIR indicate the further detection of zinc carboxylates. On occasion zinc carboxylates are detected in paint colors that do not contain zinc white and have migrated from other layers. These are indicated in square brackets.

Methods of Sampling and Analysis

Fourier Transform Infrared Spectroscopy (FTIR)

Fourier Transform Infrared Spectroscopy (FTIR) was performed on microscopic samples taken from paintings using a Thermo Nicolet Nexus Spectrometer. Samples were studied using the Continuum IR Microscope attachment and MCT-A Detector with KBr window (11,700–400 cm^{-1}). A micro compression cell with diamond window was used as a sample platform. Samples were taken by scraping the surface of the material with a sharp clean scalpel blade and pressing this material directly onto the clean diamond surface. The sample was then rolled with an FTIR roller to give thin and even sample distribution. Spectra were gathered in the range 4000–550 cm^{-1} and are the sum of a 100 scans with a resolution of 4 cm^{-1}.

* X-ray Fluorescence Spectroscopy (XRF)

Energy Dispersive X-ray Fluorescence was undertaken with a Bruker AXS handheld Tracer III-V. This instrument has a Si-Pin detector and a Rhodium target X-ray tube and a 13 µm beryllium (Be) detector window. When the Tracer head is in direct contact with the sample, the X-ray beam spot is 3 × 4 mm. The software used to operate the instrument and examine the spectra is S1PXRF 3.8.27 (Bruker AXS). The paintings were analyzed in a general setting mode using 40 kV, 1.1 µA, no filter vacuum, and vacuum pressure measured at less than 4 torr.

^ Microscopy

Samples were prepared for microscopic examination in two forms. Polarizing microscopy was performed on microscopic paint scrapings that were pressed and spread onto a glass slide. A coverslip (1 mm) was then attached using Cargille meltmount adhesive with a refractive index of 1.662. These samples were examined with an Olympus Transmission and Reflectance microscope BX51 with polarizing filters. Samples were examined at 100×, 200×, and 400× magnification.

Paint samples were also prepared as cross sections by mounting into polyester blocks and cutting after drying of the resin perpendicular to the sample surface. These were examined with the Olympus microscope BX51 with reflected light. Ultraviolet light source was provided by an Olympus reflected fluorescence system U-RFL-T with ultra high vacuum mercury burner lamp.

□ Scanning Electron Microscopy (SEM) with Energy Dispersive Spectroscopy (EDS)

Samples prepared as cross section in polyester resin were placed uncoated into the Zeiss Supra 55VP Scanning Electron Microscope at 20 kV accelerating voltage in variable pressure mode. Energy dispersive X-ray spectroscopy (EDS) was performed using the Oxford Instruments INCA x-sight microanalysis system. Pressure was maintained at 0.3 torr to minimize skirt X-rays. SEM/EDS was undertaken by Dr. Richard Wuhrer at the Microstructural Analysis Unit at the University of Technology, Sydney.

◊ Derivatized Pyrolysis–Gas Chromatography–Mass Spectroscopy (py/GC-MS)

Derivatized pyrolysis–gas chromatography–mass spectroscopy was undertaken with a Frontier Lab PY-2020D double-shot pyrolyzer system equipped with a CGS-1050E carrier gas selector, SS-1010E selective sampler, and MJT-1030E microjet cryo-trap. The pyrolysis interface was maintained at 320°C. The pyrolyzer was interfaced to an Agilent Technologies 5975C inert MSD/7890A gas chromatograph–mass spectrometer. A J&W DB-5MS-UI capillary column was used for separation (30 m × 0.25 mm × 0.25 µm), with helium carrier gas set to 1 ml/min. The split injector was set to 320°C with a split ratio of 50:1 and no solvent delay. The GC oven temperature program was 40°C for 2 minutes, then ramped to 320°C at 20°C/min, followed by a 9-minute isothermal period. The MS transfer line was at 320°C, the source at 230°C, and the MS quad at 150°C. The mass spectrometer was scanned from 33 to 600 amu at a rate of 2.59 scans per second. The electron multiplier was set to the autotune value. Samples were placed into a 50 µl stainless steel Eco-cup fitted with an Eco-stick, and three microliters of a 25% methanolic solution of tetramethyl ammonium hydroxide (TMAH) were introduced for derivatization. After three minutes the cup was placed into the pyrolysis interface, where it was purged with helium for three minutes. Samples were pyrolyzed using a single-shot method at 550°C for 6 seconds. Py-GC-MS analysis was undertaken by Michael Schilling at the Getty Conservation Institute.

Appendix II

Analysis of a Selection of Nolan's Materials from the Wahroonga and Rodd Studios, with Methods of Sampling and Analysis

TABLE 3 Analysis of a Selection of Nolan's Wahroonga Studio Materials (Sidney Nolan Archive, Art Gallery of NSW)

\# FTIR
* XRF
^ Polarizing microscopy
□ SEM/EDS
◊ Meth-Prep GC-MS
~ Raman spectroscopy

Sample	Pigments[1]	Binding medium
Can of Ripolin White (MS 2008.7/263.34.35)	Zinc white (zinc carboxylates with zinc stearate/palmitate), [barites, titanium white, chalk, kaolin]#*^□	Oil#◊
Can of Ripolin Red No. 16 (MS 2008.7/263.34.33)	Barites, toluidine red (PR3)#*^	Oil#◊
Can of Ripolin BLT White No. 1 (MS 2008.7/263.34.29)	Zinc white (zinc carboxylates), [barites, chrome yellow, iron oxide]#*^□	Oil, (castor)#◊
Can of Ripolin Black No. 1105 (MS 2008.7/263.34.28)	Carbon black#*^□~	Oil, copal, colophony#◊
Can of Ripolin BLT Ultramarine No. 13 (MS 2008.7/263.34.25)	Zinc white (zinc carboxylates with zinc stearate/palmitate), copper phthalocyanine blue (PB15), [barites, chrome yellow]#*^~	Oil, copal#◊
Can of Ripolin BLT Blue Deep No. 40RD (MS 2008.7/263.34.4)	Prussian blue, barites, chrome yellow#*^□	Oil, copal, colophony#◊
Can of Ripolin BLT PE5 (blue) (MS 2008.7/263.34.24)	Zinc white (zinc carboxylates with zinc stearate/palmitate), Prussian blue, [chrome yellow]#*^	Oil, colophony#◊
Can of Ripolin BLT No. 3056 (ochre) (MS 2008.7/263.34.27)	Zinc white (zinc carboxylates), yellow ochre, chrome yellow, titanium white#*^	Oil, copal, colophony#◊
Can of Ripolin Flat White No. 501 (MS 2008.7/263.34.30)	Zinc white (zinc carboxylates with zinc stearate/palmitate), barites, titanum white (anatase), kaolin, talc, [chrome yellow, iron oxide]#*^□	Oil#
Can of Ripolin Flat Red No. 516 (MS 2008.7/263.34.9)	Barites, toluidine red (PR3), kaolin, talc#*^	Oil#
Can of Ripolin Flat Black No. 505 (MS 2008.7/263.34.11)	Bone black, kaolin, talc#*^	Oil#
Can of Ripolin Flat Canary No. 504 (MS 2008.7/263.34.15)	Chrome yellow/sulphate, kaolin, talc#*^	Oil#
Can of Ripolin Flat Lemon No 514 (MS 2008.7/263.34.20)	Chrome yellow, kaolin, talc#*^	Oil#
Can of Ripolin Flat Blue Deep No. 540RD (MS 2008.7/263.34.3)	Prussian blue, kaolin, talc#*^	Oil#
Can of Ripolin Flat Green No. 573 (MS 2008.7/263.34.21)	Zinc white (zinc stearate/palmitate), chrome yellow, carbon black, kaolin, talc#*^	Oil#

Sample	Pigments[1]	Binding medium
Can of Ripolin Flat Green Dark Yellowish No. 590 (MS 2008.7/263.34.23)	Zinc white, Prussian blue, chrome yellow, kaolin, talc#*^	Oil#
Can of Ripolin Flat Peacock Green No. 582 (MS 2008.7/263.34.22)	Zinc white, Prussian blue, chrome yellow, kaolin, talc#*^	Oil#
Can of Ripolin Flat Pink Very Light No. 565 (MS 2008.7/263.34.14)	Zinc white (zinc carboxylates with zinc stearate/palmitate), barites, titanium white (anatase), chrome yellow, chrome red, carbon black, kaolin, talc#*^~	Oil#
Can of Ripolin Flat Maroon Lake to Shade No. 1167 (MS 2008.7/263.34.8)	Alizarin (NR9), chalk#*^	Oil#
Can of Ripolin thinners (MS 2008.7/263.37.10)	None present	Oil#
Can of Ripolin paste filler (MS 2008.7/263.42.2)	Zinc white (zinc carboxylates with zinc stearate/palmitate), chalk, titanium white#*	Oil#
Can of white paint unlabeled (MS 2008.7/263.35.1)	Zinc white (zinc carboxylates), chrome yellow#	Oil#
Can of yellow paint unlabeled (MS 2008.7/263.35.2)	Chrome yellow/sulphate, zinc white (zinc stearate/palmitate)#	Oil#
Leather satchel with 12 metal pannikins filled with paint (MS 2008.7/263.33.1 a–l)	a. Pale blue: zinc white (zinc carboxylates with zinc stearate/palmitate), Prussian blue, [chrome yellow]#* b. Black: zinc carboxylates (black not identified)#* c. Blue: zinc white (zinc carboxylates with zinc stearate/palmitate), barites, copper phthalocyanine (PB15)#* d. Dark blue: Prussian blue#* e. White: zinc white (zinc carboxylates with zinc stearate/palmitate), chrome yellow#* f. Green: Prussian blue, chrome yellow#* g. Pale red: zinc white (zinc carboxylates with zinc stearate/palmitate), chrome yellow (unidentified organic red)#* h. Yellow: chrome yellow#* i. White: zinc white (zinc carboxylates with zinc stearate/palmitate), chrome yellow#* j. White: zinc white (zinc carboxylates with zinc stearate/palmitate), chrome yellow#* k. Yellow: chrome yellow/sulfate, zinc white (zinc carboxylates)#* l. Red: barites, toluidine red (PR3)#*	a–l. Oil#
Can of B.A.L.M. (Aust.) Pty. Ltd., DULUX Black 388 Brushing Line (MS 2008.7/263.34.41)	No result# (Manganese, iron, calcium, lead, zinc, barium)*	Alkyd#
Can of Euston Lead Company (Aust.) Ltd., White lead in oil (MS 2008.7/263.34.38)	Lead white#	Oil#
T. S. Reely (North Sydney) Bitumen paint (MS 2008.7/263.34.40)	Kaolin#	Mineral oil#
Lewis Berger & Sons (Aust.) Pty. Ltd., Kem-Tone paint (MS 2008.7/263.34.39)	Titanium white (rutile), zinc white, barites, kaolin, carbon black#~	No result#
Winsor & Newton (London) Transparent Colour (blue) (MS 2008.7/263.39.6)	Prussian blue#	No result#
Berger & Son liquid stain Pale Chrome (MS 2008.7/263.34.42)	Chrome yellow/sulphate#	Oil#
Five small glass bottles, some with corks, unlabeled containing yellow solids (MS 2008.7/263.37.4–7)	None present	Oil#
Clear glass bottle, unlabeled containing yellow solids (MS 2008.7/263.37.8)	None present	Oil, natural resin#
Anslie's whisky bottle, unlabeled (MS 2008.7/263.36.5)	None present	Turpentine#
Clear glass bottle with plastic screw lid, unlabeled (MS 2008.7/263.36.11)	None present	Mineral spirits#

Sample	Pigments[1]	Binding medium
Sanitarium Health Food clear glass jar, unlabeled with brown liquid (MS 2008.7/263.40.1)	None present	Cobalt naphthenate#
Sanitarium Health Food clear glass jar, unlabeled with white residue (MS 2008.7/263.35.10)	Lead white#*	No result#
Paint Recycled Vegemite jar with bright red drips (MS 2008.7/263.35.2)	Toluidine red (PR3), barites#	Oil#
Blue cup with paint residue (MS 2008.7/263.35.3)	Titanium dioxide, barites, chalk, kaolin, ultramarine blue*#	No result#
C. FRYDER & SONS recycled tobacco can (MS 2008.7/263.35.4)	Barites, zinc stearate/palmitate, titanium white#	Oil#
C. FRYDER & SONS recycled tobacco can with dried red paint (MS 2008.7/263.35.6)	Barites, toluidine red (PR3), kaolin#	No result#
C. FRYDER & SONS recycled tobacco can containing dried red and white paint on lid (MS 2008.7/263.35.7)	Barites, toluidine red (PR3), zinc stearate/palmitate, titanium dioxide#	Oil#
Wills Capstan Navy Cut recycled tobacco can with dried black paint (MS 2008.7/263.35.8)	Ivory black, kaolin#	Oil#
Wills Capstan Navy Cut recycled tobacco can with dried blue paint (MS 2008.7/263.35.9)	Prussian blue, zinc stearate/palmitate#	Oil#
Can of Kiwi Polish Kiwi glint (MS 2008.7/263.41.2)	Kaolin, titanium dioxide, barites, chalk#*	No result#
Bottle of Winsor & Newton Winton picture varnish (MS 2008.7/263.41.3)	None present	Polycyclohexanone(AW2)#
Bottle of Winsor & Newton Distilled turpentine (MS 2008.7/263.36.8)	None present	Turpentine#
Bottle labeled Fox Bros. Pure turpentine (MS 2008.7/263.36.9)	None present	Mineral spirits#
Bottle of E. Wills Methylated spirits (MS 2008.7/263.36.2)	None present	Mineral spirits#
Bottles of N. V. Talens & Zoon Stand oil linseed (MS 2008.7/263.37.1)	None present	Oil#
Bottle with cork, handwritten label "Cobalt naphthenate" (MS 2008.7/263.40.2)	(Iron, zinc, lead, barium and sulphur*)	Cobalt naphthenate#*
Glass bottle labeled Insect lotion (MS 2008.7/263.43.2)	None present	Di-methyl phthalate#
Glass jar with pasted flower collage and pink Powder (MS 2008.7/263.43.1)	Magnesium carbonate#	None present
Jar labeled Fox Bros. Yellow ochre (MS 2008.7/263.38.1)	Yellow ochre with silicates (natural)#	None present
Paper monotype with brown and blue paint layers, center removed (MS 2008.7/263.44.7)	Dark brown: kaolin, iron oxide# Red: kaolin, yellow iron oxide# Yellow: yellow iron oxide, kaolin# Blue: Prussian blue, lead white, zinc stearate/palmitate#	No result#
Lewis Berger & Sons/Sherwin-Williams & Co./Rogers Paint & Varnish Co. Surtint color chart (MS 2008.7/263.43.10)	Burnt Sienna: iron oxide, kaolin, talc*# Pale chrome: chrome yellow*# Mid green: Prussian blue, chrome yellow*# Burnt umber: iron oxide, kaolin, talc*# Ochre: chrome yellow, kaolin, talc*# Black: No result#* French ochre: yellow iron oxide*# Deep chrome: chrome yellow*# Red oxide: iron oxide*# Prussian blue: Prussian blue*#	Burnt Sienna: Oil# Pale chrome: Oil# Mid green: Alkyd# Burnt umber: Oil# Ochre: Oil# Black: Oil# French ochre: Oil# Deep chrome: Oil# Red oxide: Oil# Prussian blue: Oil#

1. Pigments in trace amounts are indicated in square brackets.

TABLE 4 GC-MS Analysis of Drying Oils and Resins in a Selection of Nolan's Ripolin Paints

Ripolin color	Palmitic/stearic acid ratio	Suberic/azelaic acid ratio	Compounds identified with markers
BLT White 1	1.3	0.4	Castor oil
BLT Ochre 3056	1.4	0.4	Copal Colophony
Red 16	1.5	0.7	None
BLT Blue Deep 40RD	1.4	0.5	Copal Colophony
BLT Blue PE5	1.5	0.4	Colophony
BLT Ultramarine 13	1.3	0.4	Copal
Black 1105	1.3	0.6	Copal Colophony
Flat White 501	1.4	0.5	None
Flat Blue Deep 540RD	1.5	0.5	None
Flat Black 505	1.4	0.3	Copal

TABLE 5 Analysis of a Selection of Paints from the Rodd Studio

\# FTIR
* XRF

Sample	Pigments (minor)	Binding medium
Red Flo, oval tin with screw lid (SNTS 38)	No result#	Ester gum#
Flo Red D, oval tin with screw lid (SNTS 39)	No result#	Ester gum#
[illegible], oval tin with screw lid (SNTS 40)	None present	Oil#
Black Flo, oval tin with screw lid (SNTS 861)	No result#	Ester gum#
Ripolin BLT Canary No. 4 (SNTS 847)	Chrome yellow/sulphate#*	Oil#
Helizarin White T, Skilbeck (SNTS 61)	Not analysed	None present
Helizarin Brown BRT, Brico Dyestuffs Chemicals (SNTS 103)	No result#	None present
Helizarin Blue..., Brico Dyestuffs Chemicals (SNTS 211)	Copper phthalocyanine (PB15)#	None present
Helizarin Black, Brico Dyestuffs Chemicals (SNTS 259)	No result#	None present
Helizarin Turquoise GT, Brico Dyestuffs Chemicals (SNTS 358)	No result#	Butylated urea formaldehyde#
Helizarin Olive Green, Brico Dyestuffs Chemicals (SNTS 359)	Copper phthalocyanine green (PG7)#	None present
Helizarin Green GG, Brico Dyestuffs Chemicals (SNTS 360)	Copper phthalocyanine green (PG7), diarylide yellow (PY16)#	None present
Helizarin Brilliant Pink BT, Brico Dyestuffs Chemicals (SNTS 536)	Quinacridone violet (PV19)#	None present
Helizarin Yellow G, Brico Dyestuffs Chemicals (SNTS 537)	Diarylide yellow (PY16)#	None present
Helizarin Brilliant Yellow GGT, Brico Dyestuffs Chemicals (SNTS 566)	Diarylide yellow (PY81)#	None present
Helizarin Orange..., Brico Dyestuffs Chemicals (SNTS 957)	Disazopyrazolone orange (PO13)*#	None present
Helizarin Brilliant Red BBT, Brico Dyestuffs Chemicals (SNTS 959)	Monoazo, naphthol red (PR146)#	None present

Sample	Pigments (minor)	Binding medium
Ripolin Express Gloss Jasmine (SNTS 143)	Arylide yellow (PY73), titanium white#*	Alkyd#
Ripolin Express Gloss Squirrel (SNTS 145)	Titanium white, mars brown, synthetic iron oxide#*	Alkyd#
Ripolin Express Gloss Black (SNTS 156)	No result (black not identified)#	Alkyd#
Ripolin Express Gloss Dahlia (SNTS 162)	Toluidine maroon (PR13)#	Alkyd#
Ripolin Express Gloss Champagne (SNTS 164)	Titanium white (yellow not identified)#*	Alkyd#
Ripolin Express Gloss Boy Blue 1461 (SNTS 446)	Titanium white (blue not identified)#*	Alkyd#
Ripolin Express Gloss Cote d'Azur 1459 (SNTS 447)	Titanium white, copper phthalocyanine (PB15)#*	Alkyd#
Ripolin Express Gloss Red 1154 (SNTS 450)	Toluidine red (PR3)#	Alkyd#
Ripolin Express Gloss Bluebird 1436 (SNTS 454)	Titanium white, copper phthalocyanine (PB15)#*	Alkyd#
Ripolin Express Gloss Avocado 1328 (SNTS 457)	No result#*	Alkyd#
Ripolin Express Gloss White (SNTS 504)	Titanium white, kaolin#*	Alkyd#
Ripolin Express Gloss Lime 1240 (SNTS 514)	Titanium dioxide#*	Alkyd#
Ripolin Express Gloss Goldfinch 1285 (SNTS 518)	No result#*	Alkyd#
Ripolin Express Gloss Trafalgar 1463 (SNTS 519)	Copper phthalocyanine (PB15)#*	Alkyd#
Ripolin Express Gloss Blue Grass 1380 (SNTS 522)	Bromated copper phthalocyanine green (PG36)#*	Alkyd#
Ripolin Aluminium (SNTS 157)	Aluminum (iron and calcium)*#	No result#
Ripolin Polyurethane Varnish Eggshell (SNTS 147)	None present	Polyurethane#
Marabu Buntlack Seidenmattspray 013 Orange (SNTS 628)	Titanium white, diarylide yellow FGC (PY97)#	Acrylic (isobutyl methacrylate)#
Marabu Buntlack Seidenmattspray 020 Lemon (SNTS 100)	Titanium white, arylamide yellow (PY3)#	Acrylic (isobutyl methacrylate)#
Marabu Buntlack Seidenmattspray 021 Medium Yellow (SNTS 614)	Titanium white, diarylide yellow FGC (PY97)#	Acrylic (isobutyl methacrylate)#
Marabu Buntlack Seidenmattspray 023 Red Orange (SNTS 952)	Titanium white, diarylide yellow FGC (PY97)#	Acrylic (isobutyl methacrylate)#
Marabu Buntlack Seidenmattspray 028 Yellow Ochre (SNTS 607)	Mars yellow (synthetic yellow oxide)#*	Acrylic (isobutyl methacrylate)#
Marabu Buntlack Seidenmattspray 030 Vermilion Light (SNTS 757)	Titanium white, azo condensation red (PR 166)#	Acrylic (isobutyl methacrylate)#
Marabu Buntlack Seidenmattspray 033 Pink (SNTS 70)	Titanium white, quinacridone red (PR122)#	Acrylic (isobutyl methacrylate)#
Marabu Buntlack Seidenmattspray 045 Dark Brown (SNTS 997)	Diarylide yellow (PY97)#	Acrylic (isobutyl methacrylate)#
Marabu Buntlack Seidenmattspray 047 Burnt Sienna (SNTS 642)	Yellow iron oxide#*	Acrylic (isobutyl methacrylate)#
Marabu Buntlack Seidenmattspray 050 Violet Red (SNTS 759)	Titanium white, quinacridone red (PR 122)#	Acrylic (isobutyl methacrylate)#
Marabu Buntlack Seidenmattspray 052 Blue Middle (SNTS 617)	French ultramarine blue, kaolin, titanium white#*	Acrylic (isobutyl methacrylate)#
Marabu Buntlack Seidenmattspray 054 Cobalt Blue (SNTS 619)	Titanium white, barites (blue not identified)#*	Acrylic (isobutyl methacrylate)#
Marabu Buntlack Seidenmattspray 058 Paris Blue (SNTS 666)	Prussian blue, titanium white#*	Acrylic (isobutyl methacrylate)#

Sample	Pigments (minor)	Binding medium
Marabu Buntlack Seidenmattspray 062 Middle Green	Copper phthalocyanine green (PG7), arylamide yellow (PY3), titanium white#*	Acrylic (isobutyl methacrylate)#
Marabu Buntlack Seidenmattspray 067 Sap Green (SNTS 774)	Copper phthalocyanine green (PG7), arylamide yellow (PY3), titanium white#*	Acrylic (isobutyl methacrylate)#
Marabu Buntlack Seidenmattspray 065 Olive Green (SNTS 669)	No result#*	Acrylic (isobutyl methacrylate)#
Marabu Buntlack Seidenmattspray 070 White (SNTS 762)	Titanium white#*	Acrylic (isobutyl methacrylate)#
Marabu Buntlack Seidenmattspray 072 Anthracite (SNTS 646)	Titanium white (black no result)#*	Acrylic (isobutyl methacrylate)#
Marabu Buntlack Seidenmattspray 073 Black (SNTS 639)	No result#*	Acrylic (isobutyl methacrylate)#
Marabu Buntlack Seidenmattspray 075 Pine Green (SNTS 75)	Arylamide yellow (PY1), (blue no result)#	Acrylic (isobutyl methacrylate)#
Marabu Buntlack Seidenmattspray 078 Light Gray (SNTS 94)	Titanium white (black no result)#	Acrylic (isobutyl methacrylate)#
Marabu Buntlack Seidenmattspray 079 Dark Gray (SNTS 74)	Titanium white (black no result)#	Acrylic (isobutyl methacrylate)#
Marabu Buntlack Fluorescospray 331 Fluorescent Red (SNTS 760)	No result#	Acrylic (isobutyl methacrylate)#
Marabu Buntlack Fluorescospray 364 Fluorescent Green (SNTS 85)	No result#	Acrylic (isobutyl methacrylate)#
Holts Metallic Flake MF 75 Yellow (SNTS 636)	Not analyzed	Alkyd#
Holts Handy Spray Gloss Paint Blue (SNTS 643)	Copper phthalocyanine (PB15), titanium white#*	Alkyd#
Frisk Lac Gloss (Pink label) (SNTS 627)	None present	Ethyl cellulose#
Frisk Lac Matt (Brown label) (SNTS 673)	None present	Acrylic (isobutyl methacrylate)#
Holts Panel Spray Clear (SNTS 78)	None present	Acrylic (isobutyl methacrylate)#
Holts Auto Spray touch-up 2-stage clear lacquer DS101 (SNTS 81)	None present	Acrylic (isobutyl methacrylate)#
Rowney Perfix (SNTS 641)	None present	Polyvinyl acetate#
U-Spray Polyurethane clear hard lacquer (SNTS 87)	None present	Urethane alkyd#
Gillac (SNTS 91)	None present	Ethyl methacrylate#
Liquin oil painting medium quick drying (SNTS 212)	None present	Alkyd#
Liquitex gel medium acrylic polymer emulsion (SNTS 303)	None present	Acrylic copolymer (ethyl acrylate/methyl methacrylate)#
Rowney alkyd gel medium (SNTS 305)	None present	Alkyd#
Win-gel thixotropic oil painting medium (SNTS 385)	None present	Alkyd#
Helizarin binder TS 125 (SNTS 661)	None present	Nitrile rubber (polybutadiene acrylonitrile)#
Helizarin reduction binder (SNTS 583)	None present	Acrylic
Dry pigment purple, unlabeled (SNTS 64)	Manganese violet (PV16)#	None present
Dry pigment crimson, unlabeled (SNTS 65)	Litho Bordeaux red (PR63)#	None present
Dry pigment blue, unlabeled (SNTS 72)	Cobalt blue	None present
Dry pigment green, unlabeled (SNTS 76)	Copper phthalocyanine green (PG7), barites#	None present

Methods of Sampling and Analysis

Fourier Transform Infrared Spectroscopy (FTIR)

Microsamples were taken from the materials either as mixed contents of wet paints that were allowed to dry on glass slides or as dry paint residues on the containers. A micro compression cell with diamond window was used as a sample platform onto which samples were pressed and rolled flat. Samples were studied using the Continuum IR Microscope attachment in transmission mode with an MCT-A Detector with KBr window (11,700–400 cm^{-1}). Spectra were gathered in the range 4000–550 cm^{-1} and are the sum of 100 scans with a resolution of 4 cm^{-1}.

* X-ray Fluorescence Spectroscopy (XRF)

Energy Dispersive X-ray Fluorescence was undertaken with a Bruker AXS handheld Tracer III-V+. The instrument has a Si-Pin detector and a Rhodium target X-ray tube with a 13 µm beryllium (Be) detector window. The software used to operate the instrument and examine the spectra is S1PXRF 3.8.27 (Bruker AXS). Most samples were examined at 40 kV and 1.1 µA, and instrument vacuum less than 0.4 torr. In order to identify trace elements, two additional protocols were undertaken on the early Ripolin paints: 15 kV and 15 µA, with a titanium filter and instrument vacuum less than 0.4 torr; and 40 kV and 1.1 µA, with a titanium/aluminium/copper without instrument vacuum.

^ Polarizing Microscopy

A small scraping of the sample was prepared for microscopic examination by pressing and spreading onto a glass slide. A coverslip (1 mm) was then attached using Cargille Meltmount adhesive with a refractive index of 1.662. These samples were examined with an Olympus Transmission microscope BX51 with polarizing filters. Samples were examined at 100×, 200×, and 400× magnification.

□ Scanning Electron Microscopy (SEM) with Energy Dispersive Spectroscopy (EDS)

Analysis on a number of Ripolin paint samples taken from the paint-outs of the Wahroonga studio contents were examined at the Microstructural Analysis Unit at the University of Technology, Sydney. Samples were prepared as cross section in polyester resin and placed uncoated into the Zeiss Supra 55VP Scanning Electron Microscope at 20 kV accelerating voltage in variable pressure mode. Energy dispersive X-ray spectroscopy (EDS) was performed using the Oxford Instruments INCA x-sight microanalysis system. Pressure was maintained at 0.3 torr to minimize skirt X-rays. SEM/EDS was undertaken by Dr. Richard Wuhrer at the University of Technology, Sydney.

◊ Oil Analysis Using Gas Chromatography/ Mass Spectroscopy (GC-MS) with Meth-Prep II Procedure

GC-MS oils, Meth-Prep procedure: 100 µl of a solution of Meth-Prep II reagent (0.2 MTMTFTH in methanol) in toluene (1:2) was added to the vials. The vials were warmed on a hotplate at 60°C for 1 hour. After cooling, the vials were centrifuged, and the contents were ready for injection into the GC-MS. An INNOWAX (25 m × 0.2 mm × 0.2 µm) capillary column was used for the separation. Helium carrier gas was set to a linear velocity of 44 cm/sec. Splitless injection, set to 260°C, was used with a 60-second purge-off time. The MS transfer line was set to 260°C. The GC oven temperature program was: 80°C for 2 min; 10°C/min to 260°C; isothermal for 15 minutes; 20°C/min to 260°C; isothermal for 2 minutes. with a total run time of 38 minutes. The mass spectrometer was used in SCAN mode, as this permitted unknown peaks to be identified on the basis of their mass spectra using a spectral library NIST. Fatty acids were quantified with calibration curves and the amounts of fatty acids are reported as % fatty acids (mg fatty acids/mg paint). Meth-Prep II GC-MS was undertaken by Joy Mazurek at the Getty Conservation Institute.

~ Raman Spectroscopy

Raman spectroscopy was undertaken with a Renishaw inVia Raman Microscope Spectrometer with a near infrared enhanced, deep depletion charge-couple device (CCD) array detector with 576 × 384 pixels and Peltier cooling, and Renishaw near infrared diode laser at 785 nm with 300 mW nominal power at the source. This was fitted with 785 nm Rayleigh edge filters and a 1200 lines per mm dispersive grating. The microscope was a Leica DM2500M microscope with 50× reflected light illuminated objectives. Laser power at the sample was reduced to 10% using neutral density filters and pinhole inserted to improve beam. Data was collected with an extended scan and 10-second dwell time. Raman spectroscopy was undertaken by Ying Yu at the Australian Institute for Bioengineering and Nanotechnology at the University of Queensland, Brisbane.

Appendix III

Works of Art Sampled

Works not cross-referenced as figures are not reproduced in this volume.

Untitled (Abstract), c. 1939 (170.2013)
Oil on gesso on blotting paper
44.5 × 57.4 cm (17½ × 22⅝ in.)
Art Gallery of NSW, Sydney

Untitled (Abstract), 1939 (172.2013)
Oil on gesso on blotting paper
44.5 × 57.0 cm (17½ × 22½ in.)
Art Gallery of NSW, Sydney
Fig. 4

(Untitled-Abstract), c. 1939 (132.2006)
Oil on silver gelatin photograph
28.0 × 34.4 cm (11 × 13½ in.)
Art Gallery of NSW, Sydney
Fig. 9

Head of Rimbaud, 1938–39 (1980.66)
Oil, boot polish, and pencil on cardboard
34.3 × 26.9 cm (13½ × 10⅝ in.)
Heide Museum of Modern Art, Melbourne
Fig. 7

Head of a Woman (Luna Park), Sept. 1940 (166.2013)
Oil and resin on cardboard
25.0 × 20.5 cm (9⅞ × 8¼ in.)
Art Gallery of NSW, Sydney

Luna Park, 1941 (35.2003)
Nitrocellulose lacquer on canvas
67.0 × 84.0 cm (26⅜ × 33⅛ in.)
Art Gallery of NSW, Sydney
Fig. 11

Bird, Dec. 4, 1941 (1982.231)
Nitrocellulose lacquer on cardboard
45.5 × 58.5 cm (17⅞ × 23 in.)
Heide Museum of Modern Art, Melbourne

Untitled (Figure and Bird) (verso), Dec. 11, 1941 (1977.23)
Nitrocellulose lacquer on slate
25.7 × 50.8 cm (10¼ × 20 in.)
University of Queensland Art Museum, Brisbane
Fig. 10

Untitled (Round Tree), c. 1941 (167.2013)
Nitrocellulose lacquer on plywood
38.1 × 25.4 cm (15 × 10 in.)
Art Gallery of NSW, Sydney
Fig. 12

Landscape, Aug. 29, 1942 (180.2013)
Alkyd and oil on ivory wove paper
24.6 × 26.9 cm (9¾ × 10⅝ in.)
Art Gallery NSW, Sydney

On the Beach, Oct. 28, 1942 (182.2013)
Alkyd and oil on ivory wove paper
24.7 × 26.9 cm (9¾ × 10⅝ in.)
Art Gallery of NSW, Sydney
Fig. 19

Nostalgia for the Sky, Nov. 1, 1942 (183.2013)
Alkyd and oil on ivory wove paper
24.6 × 27.1 cm (9¾ × 10¾ in.)
Art Gallery of NSW, Sydney
Fig. 20

Farmer, Dimboola, Nov 2., 1942 (184.2013)
Alkyd on ivory wove paper
24.7 × 27.3 cm (9¾ × 10¾ in.)
Art Gallery of NSW, Sydney
Fig. 21

Dimboola, Dec. 5, 1942 (1982.232)
Alkyd on cardboard
62.0 × 74.7 cm (12⅜ × 29⅜ in.)
Heide Museum of Modern Art, Melbourne
Fig. 22

Water Wheel, Luna Park, 1942 (1992.112)
Alkyd on Masonite hardboard
61.0 × 92.0 cm (24 × 36¼ in.)
Heide Museum of Modern Art, Sydney

Wimmera Landscape, 1943 (1980.70)
Ripolin oil enamel on Masonite hardboard
60.8 × 91.4 cm (24 × 36 in.)
Heide Museum of Modern Art, Melbourne
Fig. 15

Self Portrait, Mar. 1943 (413.1997)
Ripolin oil enamel on hessian canvas
61.0 × 52.0 cm (24 × 20½ in.)
Art Gallery of NSW, Sydney
Fig. 25

Bathers, Apr. 12, 1943 (1982.247)
Ripolin oil enamel on cotton canvas
62.9 × 75.5 cm (24¾ × 29¾ in.)
Art Gallery of NSW, Sydney
Fig. 24

Untitled (St. Kilda), 1943 (168.2013)
Ripolin oil enamel on cardboard
31.4 × 39.5 cm (12⅜ × 15½ in.)
Art Gallery of NSW, Sydney

Boy in Township, 1943 (OA24.1962)
Ripolin oil enamel on cardboard
61.0 × 72.5 cm (24 × 28½ in.)
Art Gallery of NSW, Sydney
Fig. 26

Ned Kelly: "Nobody knows anything about my case but myself," 1945 (1988.34)
Ripolin oil enamel on cardboard
63.0 × 75.0 cm (24¾ × 29½ in.)
Heide Museum of Modern Art, Melbourne
Fig. 32

The Camp, 1946 (207.1978)
Ripolin oil enamel and alkyd
on Masonite hardboard
89.7 × 121.5 cm (35¼ × 47⅞ in.)
Art Gallery of NSW, Sydney
Fig. 31

First-Class Marksman, Dec. 12, 1946 (62.2010)
Ripolin oil enamel and alkyd
on Masonite hardboard
90.2 × 121.2 cm (35½ × 47¾ in.)
Art Gallery of NSW, Sydney
Fig. 30

Colonial Head, Apr. 4, 1947 (7.2001)
Ripolin oil enamel on cardboard
76.2 × 65.5 cm (30 × 25¾ in.)
Art Gallery of NSW, Sydney

Mrs. Fraser (originally titled *Woongoolbver Creek*), 1947 (1995.134)
Ripolin oil enamel on Masonite hardboard
65.8 × 106.6 cm (26 × 42 in.)
Queensland Art Gallery and Gallery
of Modern Art, Brisbane
Fig. 53

Island, 1947 (8.2001)
Ripolin oil enamel on Masonite hardboard
77.0 × 105.0 cm (30¼ × 41⅜ in.)
Art Gallery of NSW, Sydney
Fig. 51

Platypus Bay, Fraser Island, Oct. 1947
(2013.154)
Ripolin oil enamel on Masonite hardboard
77.3 × 106.4 cm (30½ × 41⅞ in.)
Queensland Art Gallery and Gallery
of Modern Art, Brisbane

Lake Wabby, Oct. 1947 (1982.258)
Ripolin oil enamel on Masonite hardboard
77.3 × 106.3 cm (30½ × 41⅞ in.)
Heide Museum of Modern Art, Melbourne
Fig. 52

Indian Head (Remnant), Oct. 1947
Ripolin oil enamel on Masonite hardboard
27.6 × 106.4 cm (10⅞ × 41⅞ in.)
Private collection
Fig. 56

Burke and Wills, "Grey Sick," Nov. 30, 1949
(445.1995)
Ripolin oil enamel and oil on Masonite
hardboard
91.5 × 121.5 cm (36 × 47⅞ in.)
Art Gallery of NSW, Sydney
Fig. 73

Animal in Swamp, 1958 (93.2013)
Polyvinyl acetate on Masonite hardboard
151.5 × 124.0 cm (59⅝ × 48¾ in.)
Art Gallery of NSW, Sydney

The Galaxy, 1957–58 (57.1974)
Polyvinyl acetate on canvas
193.0 × 256.0 cm (76 × 100¾ in.)
Art Gallery of NSW, Sydney
Fig. 74

Burke, 1962 (OA1.1968)
Polyvinyl acetate and oil on hardboard
122.0 × 122.0 cm (48 × 48 in.)
Art Gallery of NSW, Sydney

Elephant, 1963 (438.1995)
Oil on hardboard
152.3 × 122.0 cm (60 × 48 in.)
Art Gallery of NSW, Sydney

Mt. Erebus. Aug. 28, 1964 (1.2011)
Oil and alkyd gel medium on hardboard
122.0 × 123.0 cm (48 × 48½ in.)
Art Gallery of NSW, Sydney

Notes

Introduction

1 Brand names are typically accompanied by either ®, indicating a currently registered trademark, or ™, indicating an unregistered trademark. Henceforth in this volume the first use of a brand name will be accompanied by the appropriate trademark symbol.

2 Whether Nolan actually attended any of the night classes he was purported to have taken at the National Gallery School in 1934 is unclear. See Underhill 2015, 39.

Chapter 1

1 Chemically, the natural ultramarine blue extracted from the blue stone lapis lazuli is similar to the synthetic form, manufactured from the late 1820s. The synthetic version is distinguished by the naming convention, "French" ultramarine.

2 The FTIR absorbance peak corresponding to the carbonyl group (C=O) at 1700–1760 cm^{-1} wavenumbers is associated with the organic material in the paint samples, typically the binder. When this peak is weak or absent it can indicate that little binder is present in the sample.

3 The first plastic paint manufactured specifically for artist use was Magna™ acrylic from 1947 manufactured by Bocour Artist Colors Inc. in the USA. Liquitex acrylic was the first artist's water-borne emulsion paint available from 1954 (Learner 2004, 3).

4 Removal of original material from an artwork is always considered in light of the potential value of the information to be obtained and the condition of the artwork. Samples for paint analysis are typically taken from the edges of preexisting losses or damages and are less than 100 microns in diameter when flattened for analysis.

5 Nolan knew of Siqueiros, though precisely when and how he became aware of him is unclear. In 1961 Nolan was one of a number of prominent artists who signed a petition published in the *New York Times* (August 9) calling for Siqueiros's release after he was jailed for criticizing the president of Mexico and leading antigovernment protests.

Chapter 2

1 Sunday Reed's diaries for 1942 and 1943 are archived at JSRP, SLV, MS 13186, Box 6/18 File 9 (1942) and Box 8/File 9 (1943). The correspondence between Nolan and Sunday Reed is archived in the John and Sunday Reed Papers in the State Library of Victoria, Melbourne; for locations of individual letters cited in what follows, see the "Primary Sources" section of the reference list. Grateful acknowledgment to Dr. Richard Haese, trustee of the Reed Estate, for permission to reprint excerpts from the correspondence in this book.

2 Citizen's Military Forces Personnel Dossiers 1940–1947, B884: Record for Sidney Robert Nolan, V206559, 1939–1948, National Archives of Australia, Canberra.

3 Nolan himself seemed uncertain how to describe Dulux here, noting that it was either a lacquer or an enamel. In fact, it was neither.

4 Nolan's note about having previously prepared priming paint with egg mixed with titanium and zinc whites (tempera) is intriguing, but not yet substantiated through the analysis of paintings.

5 Traditional painting canvases are stretched over a wooden frame called a stretcher. Stretchers have unfixed corners that allow for tensioning of the canvas via the use of wooden wedges that are knocked into the corners. Strainers are a type of stretching frame that has fixed corners and cannot therefore be tensioned.

6 Jane Clark (1987) suggests that Albert Tucker may have advised Nolan in the use of marouflage, the adhering of fabric to a solid board for painting, as outlined in the technical guidebook by Max Doerner, *The Materials of the Artist and Their Use in Paintings, with Notes on the Techniques of the Old Masters* (1934).

7 A reference to John Redmond (1856–1918), an Irish politician and advocate of Irish Home Rule who backed the British (and Irish) entry into the First World War, thereby alienating many of his fellow Irish nationalists.

8 *The Herald* (Melbourne) was a broadsheet newspaper published from 1840 to 1990.

9 Reed kept two diaries for 1943; they are archived at JSRP, MS 13186, Box 8/File 9. Her entries confuse the exact dates when it comes to Nolan's shift to Ripolin paint in the first few months of 1943. In her entry of January 8, 1943, Sunday notes: "received first Ripolin Nhill dated 43 canvas 23" × 25"." However, Nolan did not write to say he had received the Ripolin until January 31 (Nolan to Sunday Reed, Jan. 31, 1943). Thus, it seems this diary entry is incorrect. A second diary kept by Reed for 1943 offers a more rational timeline. Here she recorded telephone conversations with Nolan, the letters she received from him, and his leave and transfers, in addition to some paintings as they were received. On February 2 she writes, "painted first Ripolin canvas," and on February 5, "First Ripolin painting arrived." This diary's entries are consistent with the letters, supporting the premise that the

Ripolin paint arrived on January 31, 1943, and he started painting with it just two days later.

10 Nolan himself noted when seeing this painting in 1962 that it was "originally the other way up & showed a bush or grass fire at Dimboola; later it was renamed & the St. Kilda boy added" (Art Gallery of NSW curatorial card for *Boy in Township*, handwritten note dated 10.12.62 in curator Daniel Thomas's hand).

11 Sunday Reed, 1942 diary, JSRP, MS 13186, Box 6/File 9, entries of August 19, July 26, and August 4, respectively.

Chapter 3

1 For archival information on individual letters from Nolan to Sunday Reed, see the "Primary Sources" section of the reference list.

2 After leaving Heide in 1947 for a painting trip to Queensland (see chapter 4), Nolan wrote to John Reed (Aug. 14, 1947) and asked him to send black pigment—which suggests that he had been using dry pigments and probably mixing his own paints. Analytically, it is not possible to know if Nolan added stand oil to the Ripolin paints because Ripolin itself was made from stand oil.

3 Black-and-white photographs of some of the Kelly paintings are present in the archive at The Rodd, UK (uncatalogued); in these, the works are in pale wooden frames with two steps, similar to the frames Nolan used for the Fraser Island paintings from 1947 (see chapter 4). These may be the original tulip oak frames referred to by Reeder, but close inspection of the photographs suggests that all the paintings have been temporarily propped inside the same frame, for the purposes of photography only. Nolan was not in Melbourne during 1948, and the arrangements for the exhibition were organized by John and Sunday Reed. In a letter written to John Reed from Sydney on February 26, 1948 (JSRP, MS 13186/Box 2/File 21, SLV), Nolan noted regarding the Velasquez Gallery exhibition: "As long as the frames are uniform & similar in proportion to the old ones they should work alright. The main thing is that the wood should not be too dark but you say it is only slightly more so."

4 For a discussion of the provenance and sale of this painting, see David Rainey, "*First Class Marksman:* Taking Aim at Provenance," http://www.acomment.com.au/first-class-marksman-taking-aim-provenance/.

5 *The Evening* (now in the National Gallery of Australia) was painted on the back of *Ferris Wheel* (Heide Museum of Modern Art), both Ripolin oil enamel on plywood, 90.5 × 120.0 cm (35⅝ × 47¼ in.). The splitting of the painting support was arranged by James Mollison, director of the National Gallery of Australia, in order for Sunday Reed to retain the painting *Ferris Wheel* after gifting *The Evening*, which was part of the Velasquez Gallery exhibition group of twenty-seven paintings, to the National Gallery. The treatment report describes facing the work with adhesive and tissue paper, clamping it to a table, and sawing widthwise through the middle layer of the three-ply panel (letter from James Mollison to John Reed, Dec. 14, 1978; and Ronald Earle treatment report, JSRP, MS 13186/Box 3/File 4, SLV).

6 Warwick Reeder (1997, 11) describes Albert Tucker's recollection of Nolan painting Kellys in conversation.

7 In 2007, Nolan's daughter Jinx packed up the contents of the Wahroonga studio and gifted a large collection of paints, solvents, and other assorted materials to the Artists' Materials Archive at the Art Gallery of NSW (Sidney Nolan archive, MS 2008.7/263/30-44). A smaller collection of eighteen studio items, including twelve cans of Ripolin paint, was given to the National Gallery of Victoria. Other than one can of "Special to shade No. 1135, NGA blue," the NGV Ripolin cans are replicated in the Art Gallery of NSW collection.

8 Ripolin Ltd., "Establishment of a Factory at Southall for the Manufacture of Paints and Varnishes 'Ripolin' by Dutch Company," 1931–32, Records of the Board of Trade, Office of the Chief Industrial Adviser, BT 56/47, National Archives, London.

9 Comparisons of these samples demonstrated that no further wrinkling occurred over the following eighteen months.

10 This oil is similar to that prepared by Nolan at Heide by exposing the oil to sunlight in narrow-necked jars, as discussed above.

11 It was painted at fellow artist Danila Vassilief's house while Nolan was housesitting in December 1946.

12 Whether all the Nolan Kelly paintings from 1946 and 1947 were repainted with alkyd will not be known until they, too, are analytically tested, but similar disruptions to the smooth texture of the black paint are also noticeably present in *Quilting the Armour* (1947; National Gallery of Australia).

Chapter 4

1 For archival information on individual letters from Nolan to Sunday Reed, see the "Primary Sources" section of the reference list.

2 Nolan's traveling satchel weighs twenty-four pounds (11 kg) with cans almost empty. When full, it is calculated it would weigh in the region of thirty pounds (13.5 kg).

3 Although the Wahroonga studio postdates the trip to Queensland, the cans from the studio are of a prewar Ripolin paint type; either they were sent to Nolan by the Reeds after his trip to Queensland or he was able to purchase cans similar to those he had used in Melbourne, as the analysis of the contents closely matches the Ripolin paints on works from 1943 to 1947. Moreover, samples from the cans in the traveling case match the analyzed contents of the labeled studio cans, apart from the pale red, which is not represented in any extant Ripolin.

4 Barrett Reid to John Reed, July 29, 1947, in JSRP, MS 13186/Box 2A/File 17, SLV.

5 Barrett Reid to John Reed, Sept. 28, 1947, in JSRP, MS 13186/Box 4/File 2, SLV. In Nolan's (and Barrett Reid's) letters and in his original painting titles and inscriptions, "Fraser" is (mis)spelled as "Frazer."

6 Nolan to John Reed, July 24, 1947, 4, in JSRP, MS 13186/Box 2/File 6/Part 2, SLV.

7 The Carnarvon Gorge paintings, with their stenciled hands and hunting sticks, must have deeply impressed Nolan. No doubt he recognized, through his own experience with spray paint and stencils, the power of the negative image produced in this way. Later, in 1962, he featured handprints on the costumes he made for the Royal Opera production of *The Rite of Spring;* a photograph shows stencils pinned to a dancer's costume onto which Nolan is blowing paint with an aspirator, a technique not unlike that employed by the artists at Carnarvon (Daniels 2017, 83). In 1986, he again strongly evoked the Carnarvon rock art paintings on large abstract canvases that he spray-painted, using rags as stencils (see chapter 6).

8 For example, photographs of Cooktown (now in the John and Sunday Reed papers) appear to have been used as reference in his later painting *Going to Work, Rising Sun Hotel* (1948, Ripolin on hardboard, University of Western Australia Art Collection, Perth; reproduced in Pearce 2007, 141).

9 Nolan visited the island on August 14 but likely returned to Brisbane to meet another Carnarvon Gorge expedition, which left Brisbane on the twenty-second (newspaper clipping, JSRP,

MS 13186, PA1168/Box 2/File 6, SLV; and Nolan to John Reed, Aug. 22, 1947 [see next note]).

10 Letters from Nolan to John Reed 1947–1948 are cited to the date written; they are all archived in JSRP, MS 13186, SLV: July 17, 24 and Aug. 9, 1947, Box 2/File 6/Part 2; Aug. 22, 28, Dec. 5, 1947, and Aug. 14, 1948, Box 2/File 6/Part 1; July 30, 1947, Box 7/File 1A; Feb. 3, 1948, PA1168/Box 2/File 6; Feb. 26, 1948, Box 2/File 21.

11 This August 28 letter is likely misdated, as by then Nolan had departed from Fraser Island and was joining the expedition to Carnarvon Gorge. The correct date is probably August 18.

12 Reid to John Reed, July 29, 1947.

13 Barrett Reid to John Reed, Sept. 29, 1947, in JSRP, MS 13186/Box 4/File 2, SLV.

14 "The Fraser Is show seemed one way & another to have gone fairly well. . . . Enclosing the reviews & letters he [Reid] sent down. I could not get a copy of Elizabeth Webbs (Courier Mail) review, except to read, but it was quite adverse & she talked about Masonite being a valuable building material in short supply. Which sounded somewhat ironic to my ears seeing that I had salvaged what I did get from a water logged white-anted heap on the island" (Nolan to John Reed, Feb. 26, 1948).

15 In her biography of Nolan, Nancy Underhill (2015, 191) said that there was no evidence of prior use or damage of the supports.

16 Titles in parentheses are original from the 1948 catalogue, including the spelling of Woongoolbver Creek (now Wanggoolba) (Sidney Nolan, Catalogue of Fraser Island Paintings, Feb. 17–28, 1948, Moreton Galleries, Fryer Library, John Cooper Collection, UQFL60, Box 16, Envelope 5). For simplicity, *Platypus Bay, Fraser Island* is referred to as *Platypus Bay* in the rest of the chapter.

17 FTIR analysis of the white paint on the back of four of the Fraser Island paintings (*Island, Mrs. Fraser, Indian Head,* and *Platypus Bay*) gives results for the presence of chalk, kaolin, and gypsum. These same three pigments were identified in several historic packages of Westco Kalsomine in the Artists' Materials Archive, Art Gallery of NSW. The binder identified in Westco Kalsomine was bovine animal glue (MALDI-TOF-MS analysis, Straus Center for Conservation, Kirby & Khandekar 2013). The binder in the white paint on the backs of the Fraser Island paintings has not been determined.

18 *Platypus Bay* is 77.3 × 106.4 cm (30½ × 41⅞ in.); *Lake Wabby,* 77.2 × 106.2 cm (30½ × 41⅞ in.); *Island,* 77.0 × 105.0 cm (30¼ × 41⅜ in.); *Fraser Island,* 76.0 × 104.0 cm (30 × 41 in.); and *Swamp,* 65.8 × 106.6 cm (26½ × 42 in.)

19 *Lake Wabby* is the exception here; see Appendix I. This difference in paint composition is consistent with variance in the grade of Masonite used for *Lake Wabby* as well.

20 A doublet carbonyl peak in FTIR analysis may suggest the use of a poor drying oil type.

21 Commonwealth of Australia, Queensland Lines of Communication Area, Australian Military Forces, 1943–45, *Property Assets Register—Fraser Island–1/F/2,* MSS BP241/1, National Archives of Australia, Canberra.

22 Thanks to David Rainey for drawing our attention to these photographs reproduced in Harding & Morgan 2015.

23 The photographs depict #1 *Foreshore* and #11 *Woongoolbver Creek.* In both images the same floorboards are visible in front of the paintings, showing that the photographs were taken in the same location, with *Woongoolbver Creek* photographed in front, then removed to reveal *Foreshore.* The date of the photographs is unconfirmed as they were separated from the original letter, but they are likely the photos referred to in a letter from Nolan to John Reed of December 5, 1947, where Nolan notes that he has photographed works completed for his upcoming exhibition and is waiting for the prints to send to Heide. He also refers to photographs of paintings in a letter to John Reed of February 3, 1948.

Chapter 5

1 The Director of the Art Gallery of NSW, Hal Missingham, was, however, suspended by his board of trustees from making any further purchases as a result of this purchase. "In 1949 I bought two paintings by Sidney Nolan, *Carron Plains* and *Pretty Polly Mine* for sixty-four guineas and forty-five guineas respectively. Trustees were particularly horrified at these purchases, the president remarking that if I claimed them to be poetic portrayals of the Australian landscape I was sadly misinformed on the true use of the English language, and moving from the Chair a vote of censure which, after much discussion, was unsupported. But it meant that I was soon relieved of my power to purchase" (Missingham 1973, 28).

2 Art Gallery of NSW (then the National Art Gallery of New South Wales), Biography of artist form, signed Sidney Nolan and dated June 15, 1949. Australian art curatorial file on Sidney Nolan, Joanna and Edmund Capon Library and Archive, Art Gallery of NSW.

3 Nolan's *Diary for 1950* (private collection), entry for February 17.

4 Nolan may originally have also had a Pale Blue 17, which is present on the Ripolin chart dating from the 1920s, but if so it was no longer in the Wahroonga studio in 2007.

5 For example, *Burke and Wills Expedition, "Gray Sick,"* 1949; Ripolin enamel and oil-based red ochre on hardboard; see fig. 73.

6 A small number of items from the Wahroonga studio were separated from the main group of items at the time of the gift in 2007 and given to the Conservation Department at the National Gallery of Victoria.

7 In 1950, for a first, shorter visit, Nolan had packed ten crates of paintings and materials. No doubt, the packing for his more permanent move in 1953 would have been extensive. A wooden box still in place in Nolan's last studio at The Rodd in Wales contains some materials from his 1950 trip.

8 Beck-Koller had developed a form of phenolic resins for paint that was used on Ford Motor cars in the 1920s in the United States. The company was owned by two brothers, Helmuth and Otto Reichhold, who ran the business from both the US and Austria. In 1937 Otto Reichhold was killed in the crash of the *Hindenburg* airship while on his way to visit his brother. Soon afterward, Helmuth Reichhold separated the American branch from the original Austrian family business and renamed it Reichhold Chemical Inc.

9 The paint manufacturing site at Stephen Road, Botany, is still in operation today although now owned by Allnex Holdings.

10 An emulsion is the suspension of tiny droplets of a non-water-soluble material in a volume of water. Milk, in which fat droplets are suspended in water, is a natural emulsion. Synthesized emulsions were a great boon to the paint industry, as a paint binder could be suspended in water enabling easy application and cleanup in water, but when dry the paint was washable. Emulsification is typically achieved by the use of surfactants (soaps), which provide a chemical link between the water and the non-water-soluble suspension. The first emulsion paint was called Kemtone, made with oils emulsified in water, but it was the PVA and acrylic emulsions developed in the 1950s and 1960s that transformed the paint industry.

11 PVA glue is often known by the name of the most popular trade product available in different countries. In the United States it is called Elmer's glue, in Australia Aquadhere, in Italy Vinavil. Unusually, there is no similar naming convention in the UK.

12 In about 1964, both Rowney and Art Spectrum Oil Colours supplied PVA as a paint binder for artists to mix with a range of wet pigment dispersions. These were discontinued in the 1970s (Crook and Learner 2000). It is not known when Reeves marketed PVA emulsions for artists. There is also a collection of Art Spectrum pigment dispersions for use with PVA and acrylic binders in the studio, but they do not appear to have been opened.

13 Thanks to Andrew Turley for copying and noting this pamphlet for me. Nolan's personal notes, documents, and correspondence for 1965–68 can be found in the Papers of Sidney Nolan, 1912–2011, MS 10245, National Library of Australia, Canberra.

14 Stavitsky (1999) outlines an encaustic repertoire among a number of artists in New York emerging, she suggests, from Siqueiros's Experimental Workshop in 1936—that same workshop at which nitrocellulose was also developed as an artist's medium. Wax-resist works on paper by Filipino artist Alfonso Ossorio, gifted to Jackson Pollock, may have inspired Pollock's own experiments with wax-resist techniques.

15 A range of different types of drawing crayons in the Rodd studio warrant further study in relation to Nolan's works—boxes of lithographic crayons; wax crayons, including a range of water-soluble aquarelle crayons; paint sticks; and Japanese oil pastels. An excellent investigation of the makeup of these crayons and their history of production can be found in Ellis and Yeh 1998.

16 The addressee was Sidney Nolan at 61 Deodar Road, Putney, London, and the postmark indicated a shipping date of December 17, 1957. Although the contents list on the label has been partially lost, Brilliant Violet R Helizarin, Yellow G, Blue R, and Red R are all still readable. These colors are all present in the collection of small glass jars of Helizarin dyes in the studio at the Rodd.

17 Originally called drysalters, dye companies often evolved from preservative producers into chemical and dye merchants.

18 This analytical result is supported by the Chemical Abstract Service (CAS) system, a register of chemicals that nominates a unique numeric identifier for each substance. CAS identifies Helizarin binder TS as a synonym for polybutadiene acrylonitrile. In addition, a patent from 1979 identifies Helizarin as "aqueous dispersions of copolymers on the basis of acrylates (UD and NTA) or butadiene (TS) capable of independently forming a copolymer reticulate structure" (US 4154711 A May 15 1979 US 05/731,741).

Chapter 6

1 An undated photograph of the Whitehall studio shows a table with books, drawing materials, and ornaments, all of which were found in two cardboard boxes in the Rodd barn studio in 2017, suggesting this studio was packed up when the lease was canceled on the London flat and the contents moved to Wales, then never unpacked.

2 SNTS 848 Consignment receipt for Ref No: 1166 Guns Vickers .303" G.O. No. 1. MK. 1. Quantity 75 and Ref No: 1342 Magazines No. 2. MK. 1. Quantity 400. Royal Air Force to RAAF Melbourne dated 17.11.43.

3 Pearce (2007, 243) states that the Hillman car was freighted from Sydney, whereas Adams (1987, 115) says it was purchased in the UK. The presence of spare parts in the Vickers gun case brought from Australia supports the first premise.

4 FTIR analysis cannot distinguish tung oil from other oils when dry.

5 For archival information on individual letters from Nolan to Sunday Reed, see the "Primary Sources" section of the reference list.

References

Abbreviations

AGNSW	Art Gallery of New South Wales, Sydney
AICCM	Australian Institute for the Conservation of Cultural Material
DPA&NZ	*The Decorator & Painter for Australia & New Zealand*
JSRP	John and Sunday Reed Papers, 1924–1981, SLV
HMMA	Heide Museum of Modern Art, Bulleen, Victoria
QAGOMA	Queensland Art Gallery | Gallery of Modern Art
SLV	State Library of Victoria, Melbourne
UQAM	University of Queensland Art Museum, Brisbane

Primary Sources

The correspondence between Nolan and the Reeds is found in the John and Sunday Reed Papers, 1924–1981 (JSRP), State Library of Victoria, Melbourne, MS 13186. Individual letters' archival locations are as follows:

Nolan to Sunday Reed—May 4, 1942, Box 2/File 21/Part 2; July 19, Aug. 1942, Box 7/File 4A; c. Oct., Oct. 27, Dec. 1942, Jan. 12, 1944, Box 9/File 7A; Jan. 31, 1943, Box 7/File 4B; Feb. 4, Feb. 9, May 18, July 19, 1943, Box 7/File 4C; Mar. 29, c. Autumn 1943, Box 2A/File 18; July 14, 1947, Box 97/File 2; July 15, July 17, 1947, Box 2/File 6/Part 2.

Nolan to John Reed—July 24, Aug. 9, 1947, Box 2/File 6/Part 2; July 30, 1947, Box 7/File 1A; Aug. 14, Aug. 22, Aug. 28, Dec. 5, 1947, Box 2/File 6/Part 1; Feb. 3, 1948, PA1168/Box 2/File 6; Feb. 26, 1948, Box 2/File 21.

Nolan's *Diary for 1950* resides in a private collection.

Nolan's personal notes, documents, and correspondence for 1965–68 can be found in the Papers of Sidney Nolan, 1912–2011, MS 10245, National Library of Australia, Canberra.

Nolan's Wahroonga studio materials can be found at the Artists' Materials Archive, Sidney Nolan Archive 1949–1953, MS 2008.7/263/30-44, Art Gallery of NSW, Sydney.

Sunday Reed kept a diary for 1942 and two for 1943. They are archived at JSRP, SLV, MS 13186, Box 6/File 9 (1942) and Box 8/File 9 (1943).

Letters from Barrett Reid to John Reed are also in JSRP, SLV, MS 13186: July 29, 1947, Box 2A/File 17; and Sept. 28, 29, 1947, Box 4/File 2.

Secondary Sources

"A. C. Hatrick Limited." 1934. *DPA&NZ*, Nov. 1, 56.

Adams, Brian. 1987. *Sidney Nolan. Such Is Life: A Biography*. Hawthorn, Vic.: Hutchinson of Australia.

Aftalion, Fred. 2001. *A History of the International Chemical Industry*. Philadelphia: Chemical Heritage Press.

Anderson, Michele. 1987. "Barjai, Miya Studio, and Young Brisbane Artists of the 1940s: Towards a Radical Practice." BA thesis, University of Queensland, Brisbane.

Australia. Dept. of War Organization of Industry. 1943. *W.O.I.: The Department of War Organisation of Industry, What It Is and What It Does*. Melbourne: Government Printer.

Australian Government. 1942. "Commonwealth Paint Orders 1–2." *DPA&NZ*, Jan. 17, 26.

Barber, Noël. 1964. *Conversations with Painters*. London: Collins.

Behrendt, Larissa. 2016. *Finding Eliza: Power and Colonial Storytelling*. Brisbane: University of Queensland Press.

Bertrand, Loïc, Matthieu Réfrégiers, Barbara Berrie, Jean-Philippe Echard, and Mathieu Thoury. 2013. "A Multiscalar Photoluminescence Approach to Discriminate among Semiconducting Historical Zinc White Pigments." *Analyst* 38 (16): 4463–69.

Blackman, Barbara. 1988. "Barbara's Eliza." In *Constructions of Colonialism: Perspectives on Eliza Fraser's Shipwreck*, ed. Ian J. McNiven, Lynette Russell, and Kay Schaffer, 151–56. London: Leicester University Press.

Boland, K. R. 1944–45. "The Paint and Varnish Industry in War-Time." *DPA&NZ*, Dec. 17, 12–13; Jan. 17, 12–13.

Bondfield, Robyn. 1989. *Nolan's Fraser*. Brisbane: Queensland Art Gallery.

Braun, Emily, and Carol Stringari. 2015. "Materials, Process, Color." In *Alberto Burri: The Trauma of Painting* (exh. cat.), 114–260. New York: Solomon R. Guggenheim Museum.

Burke, Janine. 2005. *In the Heart Garden: Sunday Reed and Heide*. 2nd ed. North Sydney: Random House.

Carlyle, Leslie. 2011. "Can Appearances Deceive? Visual Examination of Paint Characteristics and Rheology." Paper presented at the conference "From Can to Canvas: Early Uses of House Paints by Picasso and His Contemporaries," Marseilles, May 25–26.

Carter, Anne. 2015. "Charles Blackman's House Paints: Icy Blues and Blood Reds, Lure of the Sun." In *Charles Blackman in Queensland* (exh. cat.), 35–41. Brisbane: Queensland Art Gallery/Gallery of Modern Art.

Carter, Anne, Gillian Osmond, Paula Dredge, and Bruce Leary, forthcoming. "Zinc Oxide in Oil-Based House Paint: Insights from a Paint Chemist's Notebook Dated 1949." In *Cleaning Modern Oil Paint*. Springer International.

Carter, Anne, Gillian Osmond, and Bronwyn Ormsby. 2013. "Ian Fairweather and Water-Based Emulsion House Paints in Australia 1950–1964." *AICCM Bulletin* 34 (1): 34–43.

Casadio, Francesca, Inge Fiedler, Gillian McMillan, and Veronique Sorano-Stedman. 2013. "An Investigation into Kandinsky's Use of Ripolin in His Paintings after 1930." *Journal of the American Institute for Conservation* 52 (4): 258–77.

Casadio, Francesca, and Gwénaëlle Gautier. 2011. "Picasso at Work: Making the Case for a Scientific Re-evaluation of the Materials of the Antibes Cycle." In *Picasso Express* (exh. cat.), 135–150. Antibes: Musée Picasso.

Chanin, Eileen, and Steven Miller. 2005. *Degenerates and Perverts. The 1939 Herald Exhibition of French and British Contemporary Art.* Melbourne: Miegunyah Press.

Clark, Jane. 1987. *Sidney Nolan, Landscapes and Legends.* Melbourne: National Gallery of Victoria.

———. 2017. "Every Conceivable Approach of Eye and Mind. . . ." In *Sidney Nolan* (exh. cat.), 6–10. Birmingham: IKON.

"Crane's 'the Paint Specialists' Have a Full Range of Colours and a Unique Stock of 'Ripolin,' the Motorists' Enamel." 1921. *Sydney Morning Herald,* Feb. 3.

Crook, Jo, and Tom Learner. 2000. *The Impact of Modern Paints.* London: Tate Gallery.

Daniels, Rebecca. 2017. "Sidney Nolan and the Stage: A Holiday from Painting." In *Transferences: Sidney Nolan in Britain* (exh. cat.), 71–98. Chichester: Pallant House Gallery.

Dawe, Donovan. 1950. *Skilbecks: Drysalters, 1650–1950.* London: Skilbeck Brothers Ltd.

Doerner, Max. 1934. *The Materials of the Artist and Their Use in Paintings, with Notes on the Techniques of the Old Masters.* New York: Harcourt, Brace.

"Dr. Krumbhaar's Visit." 1937. *DPA&NZ,* June 10, 400.

Dredge, Paula. 2013. "Sidney Nolan and Paint: A Study of an Artist's Use of Commercial, Ready Made Paints in Australia, 1938–53." PhD thesis, University of Melbourne.

Dredge, Paula, Michael R. Schilling, Gwénaëlle Gautier, Joy Mazurek, Tom Learner, and Richard Wuhrer. 2013. "Lifting the Lids off Ripolin: A Collection of Paint from Sidney Nolan's Studio." *Journal of the American Institute for Conservation* 52 (4): 213–26.

"Dulux Launch." 1931. *DPA&NZ,* Dec. 1, xii–xiii.

Ellis, Margaret Holben, and Brigette Yeh. 1998. "The History, Use, and Characteristics of Wax-Based Drawing Media." *The Paper Conservator* 22: 48–55.

Gautier, Gwénaëlle. 2011. "Ripolin Product Lines." In *Picasso Express* (exh. cat.), 128–29. Antibes: Musée Picasso.

Gautier, Gwénaëlle, Anikó Bezur, Kimberley Muir, Francesca Casadio, and Inge Fiedler. 2009. "Chemical Fingerprinting of Ready-Mixed House Paints of Relevance to Artistic Production in the First Half of the Twentieth Century. Part I: Inorganic and Organic Pigments." *Applied Spectroscopy* 63 (6): 597–603.

Gettens, Rutherford, and George Stout. 1966. *Painting Materials: A Short Encyclopedia.* 2nd ed. New York: Dover.

Gleeson, James. 1979. *Transcript of Interview with Albert Tucker,* May 2. National Gallery of Australia. Available at https://nga.gov.au/Research/Gleeson/pdf/Tucker.pdf.

Gottschaller, Pia. 2012. *Lucio Fontana: The Artist's Materials.* Los Angeles: Getty Conservation Institute.

Harding, Lesley, and Kendrah Morgan. 2015. *Modern Love: The Lives of John and Sunday Reed.* Melbourne: Miegunyah Press with HMMA and SLV.

Hermans, Joen J., Katrien Keune, Annalies van Loon, and Piet D. Iedema. 2015. "An Infrared Spectroscopic Study of the Nature of Zinc Carboxylates in Oil Paintings." *Journal of Analytical Spectrometry* 30: 1600–1608.

Hoppenbrouwers, Neva. 2013. "Modernism and Masonite: A Historical Review of Hardboard Painting Supports in Australia with a Case Study from the Mary Norrie Collection." Unpublished thesis for Master's by Coursework, Centre for Cultural Materials Conservation, University of Melbourne.

It Is of Eden I Was Dreaming. 1983. Film, 25 min. Director David Muir, producer Malcolm Otton. Australia & Total Holdings, Australia. State Library of New South Wales, C4415.

The "Kelly" Paintings of Sidney Nolan, 1946–47. 1948. Melbourne: Velasquez Gallery.

King, Annette, Joyce Townsend, Bronwyn Ormsby, and Gwénaëlle Gautier. 2013. "The Use of Ripolin by Picabia in *The Figleaf* (1922)." *Journal of the American Institute for Conservation* 52 (4): 246–57.

Kirby, Dan, Narayan Khandekar. 2013. "Matrix assisted laser desorption ionisation-time of flight-mass spectrometry (MALDI-TOF-MS) of Kalsomine samples for Art Gallery of NSW." Unpublished report, Straus Center for Conservation, Harvard University, 19 April.

Klepac, Lou. 2007. "Nolan, Ripolin, and the Act of Painting." In *Sidney Nolan* (exh. cat.), 79–85. Sydney: AGNSW.

Kokkori, Maria, Ken Sutherland, Jaap Boon, Francesca Casadio, and Marc Vermeulen. 2015. "Synergistic Use of Py–THM–GCMS, DTMS, and ESI–MS for the Characterization of the Organic Fraction of Modern Enamel Paints." *Heritage Science* 3: 30.

Kubik, Maria. 2006. "Looking behind Kelly's Helmet: The Methods and Materials of Sidney Nolan." *AICCM Bulletin* 30: 5–11.

Lake, Susan. 2010. *Willem de Kooning: The Artist's Materials.* Los Angeles: Getty Conservation Institute.

Lake, Susan, Eugena Ordonez, and Michael Schilling. 2004. "A Technical Investigation of Paints Used by Jackson Pollock in His Drip or Poured Paintings." In *Modern Art, New Museums,* 137–41. Bilbao: International Institute for Conservation of Historic and Artistic Works.

Learner, Tom. 2004. *Analysis of Modern Paint.* Los Angeles: Getty Conservation Institute.

Lindsay, Francis. 2017. "Sidney Nolan: The End of St. Kilda Pier." In *Sidney Nolan* (exh. cat.), 66–77. Sydney: AGNSW.

Lipsey, David, 2017. "No. 25 Peter Grimes." In *Nolan 100,* www.sidneynolantrust.org/nolan-100/25-peter-grimes-david-lipsey.

Lynn, Elwyn. 1979. *Sidney Nolan—Australia.* Sydney: Bay Books.

———. 1985. "The Painting of Ned Kelly." In *Sidney Nolan's Ned Kelly: The Ned Kelly Paintings in the Australian National Gallery and a Selection of the Artist's Sketches for the Series,* 7–11. Canberra: Australian National Gallery.

———. 1987. "Somebody's Birthday." In *"Remembrances of My Youth": Sir Sidney Nolan's 70th Birthday Exhibition, June 11–July 12* (exh. cat.). Lanyon, ACT: Nolan Gallery, unpaginated.

Mayer, Ralph. 1940. *The Artist's Handbook of Materials and Techniques.* New York: Viking. (2nd ed., 1970.)

McCaughey, Patrick, ed. 2006. *Bert and Ned: The Correspondence of Albert Tucker and Sidney Nolan.* Melbourne: Miegunyah Press with HMMA and SLV.

McGlinchey, Chris, Anny Aviram, Sandra Zetina, Elsa Arroyo, Sil Ruvalcaba, José Luis Espinosa Pesqueira, and Manuel Eduardoy. 2013. "David A. Siqueiros: His Modification of Oil- and Cellulose Nitrate–Based Paint and His Advocacy for Innovation, 1931–1949." *Journal of the American Institute for Conservation* 52 (4): 278–89.

McPhee, John. 2013. "Modernism in the Decorative Arts." In *Sydney Moderns: Art for a New World* (exh. cat.), 114–17. Sydney: AGNSW.

Meilunas, Raymond J., James G. Bentsen, and Arthur Steinberg. 1990. "Analysis of Aged Paint Binders by FTIR Spectroscopy." *Studies in Conservation* 35 (1): 33–51.

Melville, Robert. 1982. *Sidney Nolan Chinese Journey, Oct. 6–Oct. 27* (exh. cat.). London: Thomas Agnew & Sons.

Missingham, Hal. 1973. *They Get You in the End.* Sydney: Angus & Robertson.

"Must All Great Artists Suffer? Reader Responses by James Trevelyan, Taringa; Judith Wright, Rosalie; and B.B.C, Hamilton, Brisbane." 1948. *Courier Mail,* Feb. 19, 2.

Nolan, Cynthia. 1967. *Open Negative: An American Memoir.* London: Macmillan.

Nolan Gallery, ed. 1987. "List of Works." In *"Remembrances of My Youth": Sidney Nolan's 70th Birthday Exhibition, June 11–July 10.* Lanyon, ACT: Nolan Gallery.

Ormsby, Bronwyn, and Patricia Smithon. 2008. Interview with Mary Nolan. Unpublished.

Osmond, Gillian. 2012. "Zinc White: A Review of Zinc Oxide Pigment Properties and Implications for Stability in Oil-Based Paintings." *AICCM Bulletin* 33: 20–29.

"Our Hospital." 1914. *Nepean Times,* Mar. 14, 7.

"Paint Bar Where Colours Are Mixed." 1953. *DPA&NZ,* Nov. 15, 22.

Pearce, Barry. 2007. "Nolan's Parallel Universe." In *Sidney Nolan* (exh. cat.), 17–65. Sydney: AGNSW.

———. 2017. "Planet Nolan." In *Transferences: Sidney Nolan in Britain* (exh. cat.), 9–31. Chichester: Pallant House.

Pierse, Simon. 2017. "Looking in Excited Reverie: Another Look at Nolan." Essay available at http://nolancentenary.com.au/another-look-at-nolan, unpaginated.

"Prohibition and Restrictions on Paints and Brushes. Effect upon Paint Trade." 1942. *DPA&NZ,* Jan. 17, 9.

Raeburn, Michael. 2011. "Brand Ripolin." In *Picasso Express* (exh. cat.), 123–27. Antibes: Musée Picasso.

Rafferty, Anita. 2016. Transcript of interview with Anne Carter and Gillian Osmond, Feb. 18, QAGOMA Reference Library Collection.

Rainey, David. 2012. "Nolan's 'Mrs Fraser': Reconstruction and Deconstruction." Available at www.acomment.com.au/2012/04/nolans-mrs-fraser-reconstruction-and-deconstruction.

———. 2014. "Re-discovered Nolan Images Include a Second 1947 'Mrs Fraser.'" Available at www.acomment.com.au/2014/01/[illegible]-second-mrs-fraser-1947.

Rajkowski, Raymonda. 2013. "Liquitex Acrylic Emulsion Paints: Developing Baseline Data for the Material Investigation of Works of Art by Contemporary Artists." Minor thesis, Centre for Cultural Materials Conservation, University of Melbourne.

Reeder, Warwick. 1997. "Nolan at Heide." In *The Ned Kelly Paintings: Nolan at Heide 1946–47* (exh. cat.), 9–15. Melbourne: HMMA.

"Reichhold Chemicals Inc.: A New Australian Industry." 1940. *DPA&NZ,* Mar. 12, 32–33.

Reid, Barrie. 1967. "Nolan in Queensland: Some Biographical Notes on the 1947–8 Paintings." *ART and Australia* 5 (2): 446–52.

"Ripolin Celebrates Its 80th Anniversary." 1978. *Ealing Midweek Gazette,* Mar. 7, 3.

Rogala, Dawn. 2019. "Everything Old Is New Again: Revisiting a Historical Symposium on Zinc Oxide Paint Films." In *Metal Soaps in Art: Conservation and Research.* Springer International, 315–28.

Russell, Joanna E., Brian W. Singer, Justin J. Perry, and Anne Bacon. 2012. "Investigation of the Materials Found in the Studio of Francis Bacon (1909–1992)." *Studies in Conservation* 57 (4): 195–206.

Sands. 1930. *Sands' N.S.W. Directory.* Sydney: Sands.

Sayers, Andrew. 1997. "Kelly's Words, Rousseau, and Sunlight." In *The Ned Kelly Paintings: Nolan at Heide 1946–47* (exh. cat.), 8–32. Melbourne: HMMA.

Schilling, Michael, Joy Mazurek, and Tom Learner. 2007. "Studies of Modern Oil-Based Artists' Paint Media by Gas Chromatography/Mass Spectrometry." In *Modern Paints Uncovered: Proceedings from the Modern Paints Uncovered Symposium,* 129–39. Los Angeles: Getty Conservation Institute.

Schönemann, Anna, Wolfgang Frenzel, Achim Unger, and Ernst Kenndler. 2006. "An Investigation of the Fatty Acid Composition of New and Aged Tung Oil." *Studies in Conservation* 51: 99–110.

Seear, Lynne. 1998. "A Wave to Memory." In *Brought to Light, Australian Art 1850–1965* (exh. cat.), 200–206. Brisbane: Queensland Art Gallery.

Shashoua, Yvonne. 2008. *Conservation of Plastics: Materials Science, Degradation, and Preservation.* Oxford: Butterworth-Heinemann.

Sidney Nolan: Such Is Life. 1987. Film, 60 min. Director Brian Adams. National Film and Sound Archives, Canberra, 43172.

"Siqueiros' Air-Brushed Pictures in Duco." 1940. *Art News,* January, 12.

Skinner, Peter. 1984. "John Cooper and the Moreton Galleries: A Study of a Brisbane Art Dealer and Gallery Owner 1933–1950." BA thesis, University of Queensland, Brisbane.

Smith, Bernard. 1962. Interview with Sidney Nolan. In Papers of Elwyn Lynn, MS 1997.5, Box 9, Edmund and Joannna Capon Research Library, AGNSW.

Smith, Geoffrey. 2003. *Sidney Nolan: Desert and Drought, June 6–Aug. 17* (exh. cat.). Melbourne: National Gallery of Victoria.

Standeven, Harriet. 2011. *House Paints, 1900–1960: History and Use.* Los Angeles: Getty Conservation Institute.

———. 2013. "Oil-Based House Paints from 1900 to 1960: An Examination of Their History and Development, with Particular Reference to Ripolin Enamels." *Journal of the American Institute for Conservation* 52 (3): 127–39.

Stavitsky, Gail. 1999. "Waxing Poetic: Encaustic Art in America during the Twentieth Century." Catalogue essay for an exhibition at Montclair Art Museum, Montclair, NJ, premiering May 23, 1999. Available at www.tfaoi.com/aa/2aa/2aa626.htm.

Stein, Gertrude. 1933. *The Autobiography of Alice B. Toklas.* New York: Harcourt Brace.

Stirling, Joyce. 1948. "I'm Not So Artful." *Brisbane Telegraph,* Feb. 27, 4.

Stow, Randolph. 1962. *Outrider: Poems, 1956–1962.* With paintings by Sidney Nolan. London: MacDonald.

Todd, Jan. 1998. "Cars, Paint, and Chemicals: Industry Linkages and the Capture of Overseas Technology between the Wars." *Australian Economic History Review* 38 (1): [illegible]

———. 2007. *The Polymer Story. A Family's Journey in Australian Chemical Science.* Sydney: Herman Slade Raiatea Foundation.

Townsend, Joyce, Annette King, and Bronwyn Ormsby. 2016. "The Creation and Present Appearance of Francis Picabia's 'Transparencies.'" Poster presented at "Metal Soaps in Art" conference, Mar. 14–15, Amsterdam.

Turley, Andrew. 2016. *Day-by-Day Guide to the Adelaide Ladies: Fifty-Nine Works by Sidney Nolan, 27 April 1964 to 11 June 1964.* Sydney: Andrew Turley.

Underhill, Nancy, ed. 2007. *Nolan on Nolan: Sidney Nolan in His Own Words.* Camberwell, Vic.: Penguin Group Australia.

———. 2015. *Sidney Nolan: A Life.* Sydney: New South Publishing.

van den Berg, Jorrit D. J., Klaas Jan van den Berg, and Jaap J. Boon. 1999. "Chemical Changes in Curing and Ageing Oil Paints." In *12th ICOM-CC Triennial Meeting [Lyon: 29 August–3 September] Preprints,* vol. 1, 248–53.

van den Berg, Klaas Jan, Aviva Burnstock, and Michael Schilling. 2019. "Notes on Metal Soap Extenders in Modern Oil Paints: History, Use, Degradation, and Analysis." In *Metal Soaps in Art: Conservation and Research.* Springer International, 329–42.

Webb, Elizabeth. 1948. "Shock Tactics Shock." *Courier Mail,* Feb. 18.

Weise, Gitte. 2013. "En Garde: Hera Roberts and the Burdekin House Exhibition." In *Sydney Moderns: Art for a New World* (exh. cat.), 126–31. Sydney: AGNSW.

Whiteley, Wendy. 2007. "Spaces in Time: Wendy Whiteley Talks with Barry Pearce and Alec George about Places, Spaces, and Studios." In *Brett Whiteley Studio,* 112–15. Sydney: AGNSW.

Wright, Judith. 1975. "Statement from Judith Wright McKinney, 8th August 1975." In *Fraser Island Environmental Inquiry, Transcript of Proceedings, 11 June–3 October,* 389–44. Canberra: Government Printing Office.

Illustration Credits

P. iii, fig. 76: Gerti Deutsch / Getty Images
P. 2, fig. 68: A.L. Frazer, Nolan painting Unnamed ridge in front room of Wahroonga house 1949, Cooper Collection, UQFL60, Job no F19003, Fryer Library, Queensland University Library
P. 3: Photo Brian Adams
Fig. 1: Albert Tucker, Luna Park, St. Kilda, Albert Tucker Photographic Collection, Heide Museum of Modern Art & State Library of Victoria
Fig. 2: The Argus. Newspaper Collection, State Library of Victoria
Fig. 3: John Oxley Library, State Library of Queensland Neg No: 91705 CC BY 4.0
Figs. 6, 42, 79, 83, 88, 90, 98: Photo by Author
Figs. 7, 15, 16, 18, 22, 24, 27, 29, 32, 50, 52, 59–60: Courtesy of Heide Museum of Modern Art, Melbourne
Fig. 8: Hugh P. Hall, Roman Jasinsky as Icare, in Icare, The Original Ballet Russe, Australian Tour, His Majesty's Theatre, Melbourne, 1940, National Library of Australia, nla.obj-144725486
Fig. 10: Photo Carl Warner / © The University of Queensland
Figs. 13, 86: Diagram by Author
Fig. 14: Photo courtesy Charles Osborne
Fig. 17: Caroline Simpson Library & Research Collection, Sydney Living Museums
Figs. 23, 72: Mitchell Library, State Library of New South Wales "The Decorator & Painter for Australia & New Zealand" TQ079035
Figs. 33, 35: Courtesy of Heide Museum of Modern Art, Melbourne and the Australian Synchrotron
Figs. 36–40, 43–44: National Gallery of Australia, Canberra, Gift of Sunday Reed 1977 / Bridgeman Images
Fig. 41: National Gallery of Australia, Canberra / Bridgeman Images
Fig. 45: Photo Felicity Jenkins, AGNSW
Fig. 46: Image courtesy of the State Library of South Australia no BRG 216/8/1
Figs. 47–48: Photo Author & Art Gallery of NSW
Fig. 53: Queensland Art Gallery | Gallery of Modern Art, Purchased 1995 with a special allocation from the Queensland Government. Celebrating the Queensland Art Gallery's Centenary 1895–1995, © Courtesy of the Artist's Estate / www.bridgeman.co.uk / Photo Natasha Harth, QAGOMA
Fig. 54: Private Collection / Photo by Author
Fig. 55: Image courtesy of Menzies Art Brands
Fig. 56: Private Collection, Courtesy of Sotheby's Australia / Photo by Author
Figs. 57–58, 62: Image QAGOMA Conservation
Fig. 61: Papers of John and Sunday Reed, 1924–1981. Manuscripts Collection, State Library of Victoria
Fig. 63: Exhibition of Sidney Nolan's Fraser Island paintings Moreton Galleries Brisbane 17–28 February 1948, Cooper Collection, UQFL60, Job no F19003, Fryer Library, Queensland University Library
Fig. 64: Diagram by Gillian Osmond
Fig. 65: Sidney Nolan, Baobab tree, 1950. Australian Woman's Weekly, 25 July 1950. National Library of Australia
Fig. 66: The Sydney Morning Herald / Getty Images
Figs. 70–71: Photo Jenni Carter, AGNSW
Fig. 78: Photo © Zoë Dominic
Fig. 80: Philip Quirk
Figs. 81, 84: Private Collection / Photo © Agnew's, London / Bridgeman Images
Figs. 82, 85, 89, 93–95, 97: Sidney Nolan Trust
Figs. 87, 91–92, 96: Alex Ramsey & Sidney Nolan Trust
Back cover: Albert Tucker, Sidney Nolan in the Loft, Parkville, 1945. National Library of Australia. MS 10245, Papers of Sidney Nolan, Series 2, File n. 151

Index

Acknowledgments

This book is dedicated to Laurence, Ellen, and Ruth Hall, who have been endlessly supportive and traveled with me through the world of Nolan, including to the UK to share the incredible experience at The Rodd.

The initial research for this book formed part of an Australian Research Council (ARC) Linkage grant titled *The Twentieth Century in Paint,* which contributed to a number of international projects looking at developments in paint technology in the twentieth century and their use by Australian artists. The study of Nolan's paint media formed one component of this ARC research, centered with the Linkage Partner Art Gallery of New South Wales (NSW), and was funded under the Australian Postgraduate Award (Industry) (APAI) system. I wish to thank Robyn Sloggett and Nicole Tse for conceiving the ARC project, for assembling such a diverse and inspiring group of researchers, and for their supervision of my doctoral thesis. *The Twentieth Century in Paint* project brought together painting conservators, scientists, and information technology specialists from Australia, Southeast Asia, the United States, and the United Kingdom to examine specific questions regarding artistic practice in Australia and the Southeast Asia and Pacific regions. Administrating partners are the Grimwade Centre for Cultural Materials Conservation at the University of Melbourne and the Australian Institute for Bioengineering and Nanotechnology, University of Queensland. Industry linkage partners are the Art Gallery of New South Wales, Artlab Australia, Getty Conservation Institute, National Art Gallery of Malaysia, National Gallery of Victoria, Queensland Art Gallery and Gallery of Modern Art, Tate Britain, and SEAMEO-SPAFA Regional Centre for Archaeology and Fine Arts.

The Art Gallery of New South Wales's contribution to the ARC project was funded by the Art Gallery of New South Wales Conservation Benefactors. A number of people now retired from the Art Gallery of NSW, including the director Edmund Capon, the head of Curatorial Services Tony Bond, the head of Australian Art Barry Pearce, and curator Helen Campbell, were pivotal to the project beginnings. Other staff who have continued to support the project are director Michael Brand; directors of Collections Maud Page and Suhanya Raffel; head of Conservation Carolyn Murphy, and colleagues Simon Ives, Andrea Nottage, Michelle Wassall, and Céline de Courlon; curator of Asian Art Matt Cox; head of Australian Art Wayne Tunnicliffe, and curators Denise Mimmocchi and Natalie Wilson; image sales and licensing coordinator Jude Fowler Smith; head of Photography Jenni Carter and photographer Felicity Jenkins; and art gallery librarians Steven Miller, Vivian Huang, and Claire Eggleston.

Collaborators for this Nolan research have included the conservation scientists Tom Learner, Michael Schilling, and Joy Mazurek from the Modern and Contemporary Art Research Group at the Getty Conservation Institute, who undertook gas chromatography/mass spectroscopy (GC-MS) studies on a number of Ripolin and Dulux samples from the Wahroonga studio and on various paint samples from Nolan's works. The conservation scientists Gwénaëlle Gautier and Francesca Casadio at Chicago Institute for the Arts also kindly shared information regarding their own research in Ripolin paint manufactured in France and undertook with this researcher some early GC-MS studies on the Ripolin paint from Nolan's studio. Funding for travel to the United States to undertake the GC-MS analysis with these two groups of scientists was provided by the ARC *Twentieth Century in Paint* project, the University of Melbourne under its TRIPS funding.

Other scientific partners include John Drennan and Ying Yu at the Centre for Microscopy and Microanalysis, University of Queensland, who undertook Raman spectroscopy, scanning electron microscopy, and X-ray diffraction studies on a number of Ripolin samples. Gillian Osmond, Stephen Best, and Nicole Tse were all involved in the examination of Ripolin samples with the Infrared Beamline at the Australian Beam-line, together with the Synchrotron scientists Ljiljana Puskar and Mark Tobin. Australian Synchrotron scientists Daryl Howard and David Paterson undertook the XRF mapping project of the Kelly painting from the Heide Museum of Modern Art. Andrew Durham, Helen Weidenhofer, and Gillian Leahy from Artlab

Australia kindly facilitated the analysis of the Ripolin color chart in the collection of the State Library of South Australia and gave access to their instrumentation. Richard Wuhrer at the Microstructural Analysis Unit, University of Technology Sydney, undertook scanning electron microscopy analysis on a number of Ripolin paint samples and samples from paintings by Nolan.

Access to collections for analytical studies has been essential for the success of this project, but it is a special and rare privilege for an outside researcher. The Heide Museum of Modern Art generously contributed to the study group of pictures, particularly from the critical wartime and postwar years, and entered the process with me of undertaking the X-ray fluorescence mapping at the Australian Synchrotron on the painting *Ned Kelly: "Nobody knows. . . ."* Curator Kendrah Morgan has been wonderfully supportive of this work and has openly shared her knowledge of Nolan's work with me along the way. The Queensland Art Gallery and Gallery of Modern Art provided their significant Queensland paintings by Nolan as subjects for study. The University of Queensland Art Museum provided several samples from an early slate painting for analysis. Geoffrey Smith, chairman of Sotheby's Australia, provided access to a lost Fraser Island painting off-cut, and generously shared his extensive experience and knowledge of Nolan's Sydney period and the Wahroonga studio. Painting conservators John Payne and Michael Varco-Cox arranged and facilitated access to the Nolan Wahroonga studio material in the collection of the National Gallery of Victoria. Dulux History project archivist Trudy Scott and retired Dulux paint chemist Bruce Leary also made available their records, paint collections, and expertise.

Funding for a research residency at Nolan's UK studio The Rodd was provided by the Gordon Darling Foundation, Rowena Danziger and Ken Coles, the Sidney Nolan Trust, and the Art Gallery of NSW. The Sidney Nolan Trust has been welcoming of this research and generously allowed the use of images and words by Sidney Nolan for this book, along with Bridgman Art Library. Thanks are due particularly to the director Anthony Plant, public relations manager Amanda Fitzwilliams, and trustee Rebecca Daniels, who were all so warm in their welcome of my research and presence during my residency in 2017. Permission for the use of the contents of Sidney Nolan's letters to Sunday Reed was given by Richard Haese, trustee of the Reed Estate.

As my doctoral thesis morphed into this book project, many people at the Getty Conservation Institute supported its development. In particular I wish to thank Tom Learner, head of Science, who first encouraged me to join the ARC research project and then supported the expansion of my thesis into a volume for the Artist's Materials series. Cynthia Godlewski, publications manager, nurtured the process through the permissions, acceptance, and peer review steps. Thanks also to Narayan Khandekar and Simon Pierse, for providing insightful comments on the manuscript, and to the team at Getty Publications who took the project through its next steps: project editor Beatrice Hohenegger, freelancers Anne Canright for her thoughtful editing and Pamela Morgan for her meticulous proofreading, Kelly Peyton for the task of locating images and navigating permissions, Zoe Goldman for editorial assistance, Jeffrey Cohen for design, and Suzanne Watson for production.

Critical to this book project are contributors Anne Carter and Gillian Osmond, conservators at the Queensland Art Gallery and Gallery of Modern Art, who willingly joined with me in writing the chapter on Nolan's Queensland paintings. Gillian has been present through the ARC research project, lending her extraordinary intelligence and knowledge, particularly in the area of zinc carboxylates. A special acknowledgment of Jinx Nolan's generosity and foresight in gifting the Wahroonga studio contents to the Art Gallery of New South Wales must be given. The studio material has been the catalyst for the research, and this project would not have existed without this inspirational gift.

Paula Dredge

About the Authors

PAULA DREDGE studied conservation of cultural materials at the University of Canberra and art history at the University of Sydney. She received her PhD from the Grimwade Centre for Cultural Materials Conservation at the University of Melbourne. She is the head of paintings conservation at the Art Gallery of New South Wales (NSW), where she has worked since 1990. In 2017 she was awarded the Australian Institute for Conservation of Cultural Materials award for outstanding research in the field of conservation, and in 2018 she was part of the team awarded the Museums & Galleries National Award (MAGNA) for interpretation, learning, and audience engagement (level 1) for the *Sidney Nolan Unmasked* exhibition at Heide Museum of Modern Art, an augmented reality experience using the Synchrotron XRF mapping of *Ned Kelly: "Nobody knows anything about my case but myself."*

ANNE CARTER is a paintings conservator at the Queensland Art Gallery | Gallery of Modern Art in Brisbane. After completing degrees in Studio Art (Illustration) and Art History, Anne earned a Bachelor of Applied Science in Conservation of Cultural Materials (Paintings) degree from the University of Canberra. She later completed a two-year internship at the Museum of Modern Art, New York, focusing on devarnishing modern paintings. Anne continues to gain experience in the conservation of modern and contemporary paintings and has particular interest in the social history of artists and their materials. She has published on the paintings of Australian artists Ian Fairweather and Charles Blackman and their relationship with the Australian house paint industry.

GILLIAN OSMOND is a paintings conservator at the Queensland Art Gallery | Gallery of Modern Art in Australia with three decades of industry experience. She has a Bachelor of Applied Science in Conservation of Cultural Materials (Paintings) degree from the University of Canberra. In 1992 she completed a research internship at the Tate Gallery in London, developing a longstanding interest in the technical examination of paintings and paint cross-sections. In 2014, Gillian earned a PhD from the University of Queensland for her research on zinc oxide–centered deterioration in artist's oil paints and consequences for paintings.